SWARMING
ITS CONTROL AND PREVENTION

Northern Bee Books

SWARMING AND ITS CONTROL AND PREVENTION
© L. E. Snelgrove

This reprint is of the Seventeenth Edition October 2014

Northern Bee Books
Scout Bottom Farm
Mytholmroyd
Hebden Bridge
HX7 5JS (UK)

ISBN 978-1-914934-00-1

Front Cover © John Phipps

D&P Design and Print
Worcestershire

SWARMING
ITS CONTROL AND PREVENTION

Northern Bee Books

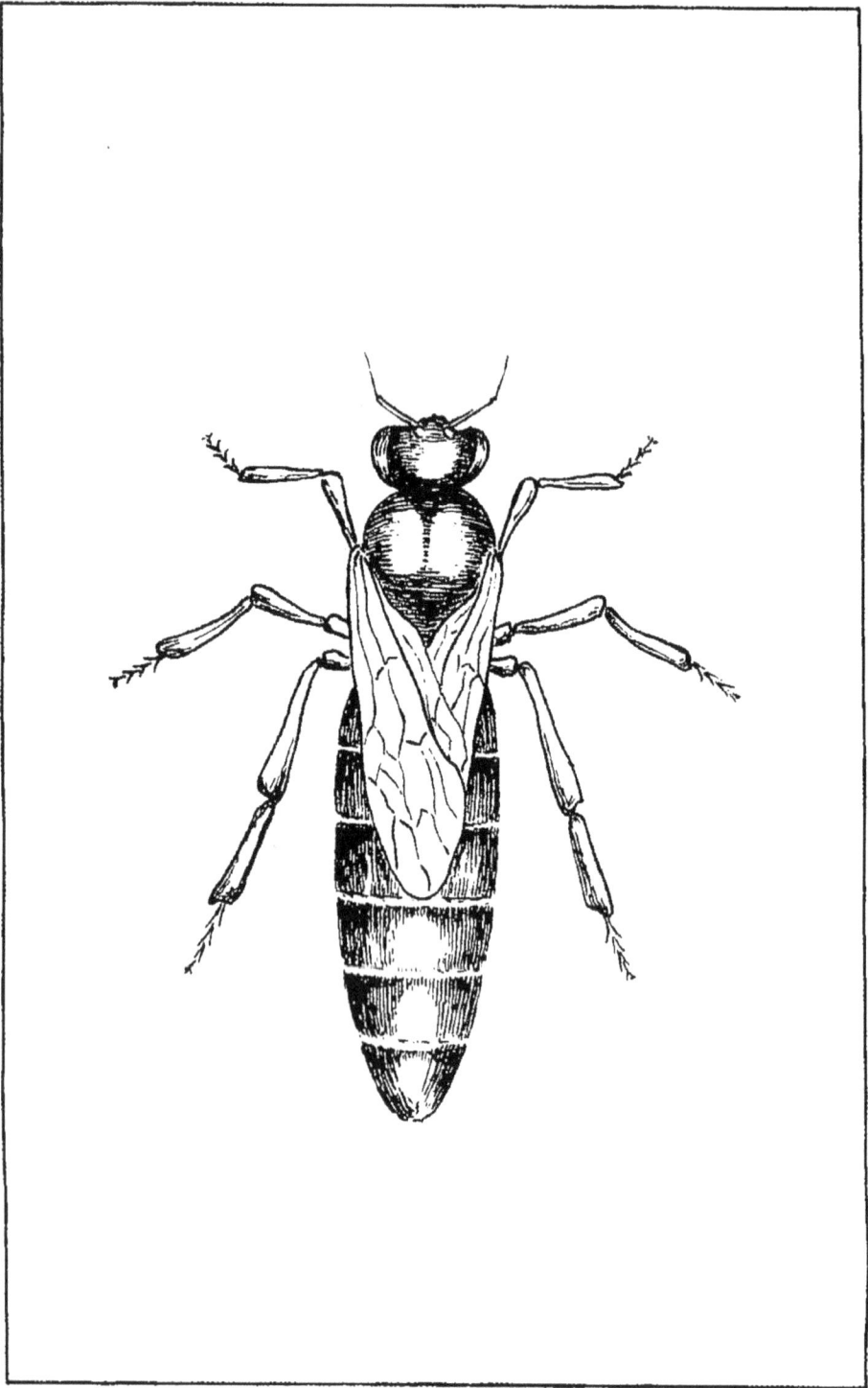

Queen Bee and illustrations throughout the text—Irene Snelgrove

SWARMING
ITS CONTROL AND PREVENTION

By

L. E. SNELGROVE, M.A., M.SC.

Fellow of the Royal Entomological Society
President of the Somerset Beekeepers' Association
Hon. Life Member of the British Beekeepers' Association
Expert and Honours Lecturer, British Beekeepers' Association
Past President (1956) of the British Beekeepers' Association

PREFACE TO THE FIRST EDITION

THE problem of controlling and preventing the swarm-ing of bees has exercised men's minds ever since bees were first kept in hives. The invention of movable combs, which greatly increased the possibility of control, in some respects intensified the problem. Most beekeepers of to-day have little difficulty in managing bees successfully during ten months of the year but when the swarming months come and they are anticipating a good yield of honey as the reward of their labour and care, they are dismayed and often reduced to a state of despair by the frequent dis-appearance of swarms from their best stocks. Indeed, apart from bee-diseases, the difficulty of swarm-control still remains the greatest obstacle to successful bee-keeping. Many methods of swarm-prevention have been devised but have not been widely adopted,—some because they involve difficult, frequent, or un-pleasant manipulations; others because they involve too much labour, are unreliable, or involve an element of cruelty; and others because they interfere with the main purpose of keeping bees,—that of honey pro-duction.

Having been obliged to keep bees in out-apiaries I have devoted the past ten years to the study of swarm-control. During that time I have discarded one by one, as being laborious, difficult, uncertain, or unprofitable, the best known methods of prevention described in bee literature, as well as original ones of my own. My aim was to be able to leave bees in an out-apiary with-out constant anxiety and fear of loss during the swarm-ing season and succeeding honey-flow. During the winter of 1930–31 I devised a plan which I applied successfully to 16 hives. In September 1931 I was persuaded, prematurely, to lecture on this plan at the Seale-Hayne Agricultural College. In the following year I practised it on 23 stocks, 22 of which did not swarm, and again lectured on it at the Seale-Hayne College in September, and also to the Somerset Bee-keepers' Association at Taunton in February. These two seasons were poor ones for honey production. In 1933, a good season, I applied the method to 34 stocks,— again without swarms except in one case. Curiously enough I had one swarm in each year. In 1931 a stock threw a swarm late in August,—long after the honey season was over. In 1932 one came out on the last day in July. There was not a cell of honey in this hive and I concluded it was a "starvation" swarm. In 1933 one stock swarmed the day after the method was applied to it, probably on account of omission to provide it with a little unsealed brood.

As a result of my lectures at several meetings of beekeepers I have been urged by many to give a detailed written description of the method, both as applied to the simple hive I use myself and, in a modified form, to any ordinary bar-frame hive.

It was only last year that I discovered how to apply it so as to prevent swarming from stocks with advanced queen cells without depriving such stocks of their queens or destroying the queen cells. This, I believe, has hitherto been considered impossible.

I claim that my method is simple and original and that it has the following advantages:—

(1) It involves the expenditure of very little time and labour.

(2) It does not interrupt honey gathering nor the queen in her egg-laying.

(3) A minimum of equipment is needed.

(4) There is no need to catch the queen if it is desired not to do so, or to break down queen cells.

(5) It provides for increase or not, as the beekeeper may desire.

(6) It provides for annual re-queening and selection of stock.

(7) It ensures immediate occupation of supers, whether containing sections or extracting combs.

(8) It ensures a full honey crop.

(9) It enables the beekeeper to send exceedingly strong stocks to the heather.

I do not pretend that the method will ensure increased surplus honey or that the bees will gather surplus in a bad season, but I may quote as an example of its efficacy the case of a hive which I worked last year in the garden of my friend, Mr. A. Pilkington, of Milborne Port, 50 miles from my home, which yielded a total of 230 lbs. of honey together with three strong nuclei headed by young queens.

I have considered it desirable to describe and criticise in the later pages of this book several other methods of swarm control, including one of my own, and also to make some observations on various methods of artificial swarming, and the utilisation to the best advantage of a natural swarm.

I am indebted to several friends for help in the testing of my method and particularly to Mr. F. W. Moore, of Clevedon, who kindly prepared the diagrams used in this book, and to Dr. J. Wallace who has taken an interest in my experiments and revised my manuscript.

L. E. SNELGROVE

Bleadon,
    Weston-super-Mare.
        March, 1934.

INTRODUCTION
by Karl Showler, President BBKA 1989–1990

I WAS pleased to respond to John Kinross' invitation to write a preface to BBNO's reprinting of L.E. Snelgrove's *Swarming* as Betty and I are members of that steadily shrinking group who heard Snelgrove's last lectures in the 1950s.

Swarming its Control and Prevention is the only monograph known to me on these two important interlinked subjects although most practical beekeeping books deal with swarms and swarm control to a greater or lesser extent. No other book covers in such a wide ranging way the practicalities of swarming and swarm control.

Technically it will be of benefit to beekeepers generally to keep the book in print although our understanding of the mechanisms of swarming have advanced over since *Swarming* was first published in 1934. Dr Mark Winston in his *The Biology of the Honey Bee*, (Harvard, 1987) brings us up-to-date on current knowledge and leads into the topic that did not feature in Snelgrove's book: 'queen supersedure' i.e. the natural replacement of the queen by the colony. Many beekeepers fail to appreciate that all queens have to be replaced either by the bees through her supersedure by a daughter or by the beekeeper through the introduction of a new queen. Beowulf Cooper's *The Honeybees of the British Isles*, (British Isles Bee Breeders Association, 1986) deals with supersedure at length.

Swarming falls into two principal parts of the first deals with the ingenious board devised by the author, who developed the ideas current in the 1920s. By the turn of the 19th century as the use of modern hives became common, 'swarmboards' were devised, for example J.E. Chamber's, Vigo, Texas, described a similar swarm board to that of Snelgroves's in the US magazine *Bee Culture*, 8 November, 1906.

The second and third chapters form Part Two and are devoted to other methods of swarm control; a final chapter deals with the 'use' of swarms and the book concludes with a useful appendix on the causes of swarming.

For those who find Snelgrove's own methods complex or do not wish to get involved in double brood chamber systems there is much to learn in the second and third chapters.

Snelgrove mentions the 'Heddon' method in Chapter 2 Section 9 and Chapter 4 'Reinforcement by Heddon Method'. James Heddon was a now-forgotten US beekeeper who, in the 1880s, advocated a small brood chamber by replacing combs with dummy frames to get the bees up into the supers to produce sections. He also obtained this reduction in broodnest size by using supersized boxes for the brood chamber which could thus be reduced horizontally.

As beekeepers we have not used Snelgrove's board but have practiced many of the methods he mentions and we value this book as it brings so much together into one place.

KARL SHOWLER, Hay-on-Wye, Hereford   May, 1995

CONTENTS

PRELIMINARY CONSIDERATIONS

UNTIL comparatively recent times it was the general practice to keep bees in small hives with fixed combs. Beekeeping was a simple matter, consisting of little more than the capture and hiving of swarms as they issued, and the subsequent taking of their honey. The more numerous the swarms the more pleased their owner would be. He and his family watched diligently for them from May till July, hived them when "taken" with joyous ceremony, and then left them to take their chance during the ensuing honey season. In the autumn the heaviest hives were selected, the bees destroyed, the slabs of honey cut out, and the empty hives stored in readiness for the next season's swarms. It mattered little that some of the swarms escaped or that some of the stocks died of starvation in the winter for a single hive would usually provide several swarms in one season. The crops of honey were small, but so also were the labour and expense. The frequent renewal of queens and combs reduced the incidence of diseases, which were generally ignored. In a bad season there was little profit, but in a good one the beekeeper could provide his family and friends with honey, wax, and mead, and often had a surplus for market.

Even with such crude methods, however, it was inevitable that thoughtful beekeepers would observe that, *caeteris paribus*, swarms issued earlier and more frequently from the smaller hives and that the larger hives gave less swarms and yielded more honey. Extension of fixed-combed hives by means of "supers" tended to reduce swarming and to increase honey crops, but the possibilities of such extension were limited. The introduction of movable combs about the middle of last century facilitated the enlargement of hives with consequent increase of the size of stocks and the honey they stored. It was realised that to obtain a heavy yield of honey it was essential to secure the greatest possible bee population in a hive at the opening of the honey season, and if possible to keep this population busy at home rather than to allow it to dissipate its energies in forming new colonies by means of swarming. It was easy enough to secure the populous stock, but to prevent it from swarming was another matter,—one which has remained an insuperable difficulty for the majority of beekeepers.

SPRING DEVELOPMENT.

It has often been said that a season's work in an apiary begins at the end of the previous summer. It is then that the prudent beekeeper makes sure that his stocks are headed by young queens, provided with ample stores, and housed in clean, dry, draught-proof, and properly ventilated hives. On the approach of cold weather the queens cease to lay eggs and brood rearing is terminated for the year. The bees, clustering more or less closely, pass two or three months in restful inactivity and usually need no attention during this period. Their numbers gradually diminish and may ordinarily be considered to be at a minimum about the end of January. Soon after the "turn of the days" however, especially if some mild weather is experienced, the queen of a good colony begins to lay again. The number of eggs deposited daily in the cells is very small at first but increases little by little until it reaches a maximum in May or June. At some time in this period of six months the "births" begin to exceed the deaths in the colony which thenceforward gradually increases in strength.

SPRING STIMULATION.

It is generally known that the queen can be stimulated to lay more rapidly if the bees are fed with thin syrup. Some people stimulate their stocks too early. It is undesirable to hasten development in this way so as to secure a maximum population of field bees some time before the main honey-flow arrives. Many competent people consider spring stimulation unnecessary and even undesirable, but if it is resorted to, a little consideration will enable one to determine the most profitable period for it.

A worker bee emerges from its cell three weeks after the egg is laid. It does not usually become an active honey-gatherer until about three weeks later, that is, six weeks after the laying of the egg, and it may continue as a gatherer for about six weeks in the busy season.

Now let us assume that the usual recognised period of main honey flow in a locality comprises the last two weeks in June and the first two weeks in July, and that a bee is an active forager from her sixth till her twelfth week. In order that a bee may put in at least one week's field work during this honey flow the egg from which it is produced should be laid by the queen not later than six weeks before July 7th, i.e., 26th May, and not earlier than twelve weeks before the 23rd June, i.e., 1st April. Thus we may say that bees resulting from eggs laid from 1st April till 26th May will be active field bees during from one to four weeks of the main honey flow. Those hatched from eggs laid between 21st April and 5th May will be foragers throughout the honey flow. Spring stimulation, whether natural (as from fruit blossom) or artificial, would therefore have a maximum value during this latter period (vide diagram on page 12). Similar calculations, which are, of course, of only approximate value, can easily be made in respect of other periods of honey flow.

DIAGRAM ILLUSTRATING APPROXIMATELY THE MOST PROFITABLE TIME FOR SPRING STIMULATION IN RELATION
TO A PERIOD OF HONEY FLOW

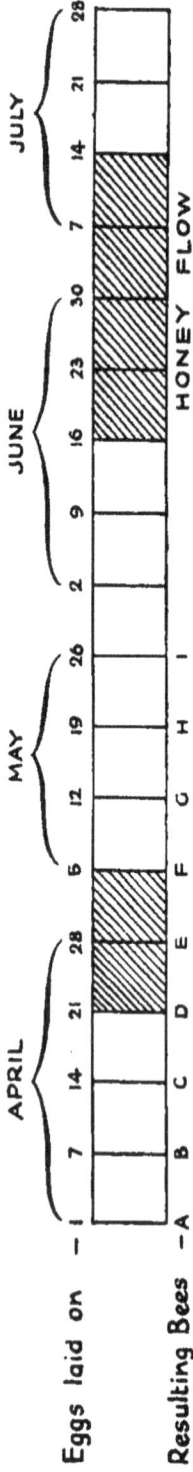

APRIL				MAY				JUNE				JULY		
7	14	21	28	5	12	19	26	2	9	16	23	30	7	14	21	28

Eggs laid on — 1

Resulting Bees —A   B   C   D   E   F   G   H   I

HONEY FLOW

Best stimulating period

A does 1 week's work during the honey flow.
B ,, 2 ,, ,, ,, ,, ,,
C ,, 3 ,, ,, ,, ,, ,,
D ,, 4 ,, ,, ,, ,, ,,
E ,, 4 ,, ,, ,, ,, ,,
F ,, 4 ,, ,, ,, ,, ,,
G ,, 3 ,, ,, ,, ,, ,,
H ,, 2 ,, ,, ,, ,, ,,
I ,, 1 ,, ,, ,, ,, ,,

Spring stimulation by feeding is of importance in seasons when weather conditions do not permit of bees gathering nectar from the fruit blossoms. In any case it is essential that all possible steps be taken to ensure a maximum force of field bees at the time of the main honey flow. It is of little use to practise non-swarming methods on backward colonies which reach their greatest strength when the honey flow is over.

YOUNG BEES AND THEIR FUNCTIONS.

During the first three weeks of their lives young bees are occupied in various domestic duties. At first they do not fly from the hive. During the first week they learn to fly and disport themselves daily in front of the hive in order to take their bearings. As a rule they do not become active foragers until they are about three weeks old.

According to Dr. Rösch (quoted by Morland, *Annals of Applied Biology*, February, 1930) the principal domestic duties of the young bees are as follow:—

(1) From the first to the third day they prepare the empty cells for the brood and assist in incubating it by maintaining the necessary hive temperature;

(2) From the third to the sixth day they feed the older larvae with pollen and honey;

(3) From the sixth to the tenth or fifteenth day they feed the younger larvae with larval food,— now generally considered to be a special secretion of the pharyngeal glands which are in a state of maximum activity in bees of this age;

(4) From the tenth to the twentieth day they are engaged in various domestic duties, including the receiving and ripening of honey and the secretion of wax.

Older bees can produce both brood food and wax provided their respective glands are not worn out and atrophied by use. There is considerable over-lapping therefore in respect of the exercise of these two functions, especially in times of comparative inactivity.

Roughly speaking we may say that the younger nurse bees are those which have not learnt to fly and that the older nurses and wax-secreting bees should be included with the flying bees. If therefore we deprive a stock of its "flying" bees we take from it most of the older nurses, wax-secreters, and the foragers.

The larval food is not only supplied to the young larvae but is also given to the queen and drones, and is provided in abundance (probably with some modification) as "royal jelly" to young queens throughout their larval stage.

For convenience I propose to use the term "nurse bees" as applying to the older nurses which would leave a hive moved to a new and distant position. The terms "foragers" and "field bees" will apply to bees over three weeks old, and "flying bees" will connote the bulk of the nurses and wax-secreting bees together with the foragers, that is to say, all the bees which would be abstracted from a hive moved to a distant position.

CONDITIONS CONDUCIVE TO SWARMING.

Amongst conditions popularly supposed to cause swarming are:—

(1) Lack of space for the extension of the brood nest;

(2) Lack of storage room for honey;

(3) Crowding of the brood nest;

(4) Insufficient ventilation;

(5) High temperature;

and to deter or prevent the swarming the beekeeper is advised to increase the capacity of the brood nest, add storage room in advance, provide shade, and ensure increased ventilation. Experience soon shows him, however, that these measures may delay, but they cannot be depended upon to prevent the issue of swarms.

It can be shown, however, that in themselves these unfavourable hive conditions do not necessarily induce a swarming impulse in the bees. Rather should they be regarded as attendant circumstances of whatever may be the true cause of swarming. As the spring advances the queen, under the stimuli of rising temperature and more liberal supplies of food presented to her by her attendants, gradually increases the rate of her laying until this reaches a peak or maximum. She may maintain this peak rate for a few days, after which her laying powers slowly decline. Since the time of emergence of the young bees from their cells lags behind that of the laying of the eggs by three weeks, it follows that a time comes when bees are "hatching" in increasing numbers while the queen is laying at a diminished rate. The result is that in the seasonal history of a normal stock there is a period when large numbers of nurse bees, eager to feed larvae, and of wax secreters ready to build new combs, can find little employment, and are therefore in excess of the hive requirements. This period coincides with the time when the foragers begin to bring in quantities of fresh nectar and pollen.

Instinctively the bees seem to realise that this is the favourable time for colonising; they quickly develop the swarming impulse and the beekeeper often finds it extremely difficult or impossible to prevent them from fulfilling their purpose.

About 40 years ago the German investigator Gerstung propounded the theory that the development of the swarming impulse is due to the presence in the hive of an excess of nurse bees and a consequent super-abundance of larval food. This theory has been widely accepted but frequently challenged. (See Chapter VII.) At all events it enables us to account for much in connection with swarming, and as a working hypothesis, which at the time I accepted as sufficient, it led me three years ago to devise the swarm-prevention method described later in this book.

There are two kinds of swarms, viz.: "Hunger" swarms and mating flight swarms, which must not be confused with ordinary swarms. To them ordinary methods of control do not apply.

Occasionally a stock of bees, reduced in numbers and faced with immediate famine, will abandon its poor home, issuing as a swarm, perhaps in the instinctive hope of making a new start in life. Generally it is use-less, refusing to re-occupy its old quarters even if honey is provided for it. It is usually met with when brood cannot be spared from another stock.

As a rule a young queen proceeds on her mating flight alone. Sometimes however her bees are tempted to accompany her. In this case they settle as a swarm, which, if not captured and hived, may be lost. Loss of queens and bees in this way is a great source of annoyance to people who use small nuclei for queen mating and little can be done to prevent it.

CONDITIONS WHICH RETARD SWARMING.

Lecturers on beekeeping often advise their listeners to breed queens only from stocks which show little or no inclination to swarm, and in this way to develop non-swarming strains. This is a counsel of perfection, —not usually a practicable one for the small beekeeper, who cannot control the drones from his neighbours' bees. Some of the larger beekeepers appear to have made some progress in this direction, but they are not thereby relieved of anxiety in the swarming season

The continued importation of foreign bees with marked swarming propensities militates against systematic selection.

It is commonly believed that stocks headed by young queens are less inclined to swarm than those with older queens. This is especially so when young queens head stocks in which they have recently been reared.

A prolific queen in a hive which is judiciously enlarged may keep her young bees busy throughout the season and therefore her stock may not swarm.

The mere re-arrangement of the combs in a hive has some inhibitory effect on swarming. If a stock is found with incipient queen cells in which the queen has already laid eggs, the otherwise inevitable swarm can be delayed for some time by crushing all the queen cells with the finger and alternating the brood combs with empty combs or sheets of foundation. This should be done only with a populous stock and in mild weather as otherwise the sudden doubling of the capacity of the brood nest may cause some chilling of brood.

Continued wet or cold weather during the swarming period sometimes causes bees to destroy queen cells they have made in preparation for swarming, which is thereby delayed for some time. The advent of a heavy honey flow likewise deters them, provided preparations for swarming are not already begun.

It is supposed by many that the presence of large numbers of drones promotes swarming. This may be true, but bees will swarm even if they are practically without a drone population. The careful beekeeper keeps drones at a minimum in hives worked for surplus, quite apart from the question of swarming.

It is well known that certain varieties of bees are more addicted to swarming than others. Generally speaking Dutch and Carniolan bees are very troublesome in this respect and, notwithstanding their other admirable qualities, should be avoided.

CHAPTER I

METHODS OF SWARM PREVENTION

In beekeeping, as in other occupations concerned with nature, it is essential to work intelligently, to be familiar with principles, and to be able to modify and adapt them as varying circumstances demand. It has often been said that "bees do nothing invariably", but they conform to certain approximate laws of life of which we make use in dealing with them. In our part of the world the vagaries of the weather present unexpected problems to an apiarist which tax to the utmost his patience and resourcefulness.

In describing the following non-swarming method therefore I propose to discuss fully the reasons for procedure, to deal with some probable contingencies, and afterwards to give simple directions for carrying out the method as a whole.

The reader should realise that although the method needs a detailed and somewhat elaborate description it is extremely simple to apply, the actual net time which should be necessary to carry it out on a stock during a season being considerably less than one hour.

On the assumption that the presence of an excess of nurse bees induces the swarming impulse I reasoned that if by some simple means the nurse bees and field bees could be separated from the others the impulse would not be developed. This has proved to be the case, and this separation is the distinguishing charac-teristic of a method which can be effected without at any time interrupting the queen in her duties or diminishing the force of field bees during the honey flow.

During 50 years of beekeeping I have had experience of most of the types of hives in common use, and have invented some in which were incorporated what were probably original ideas. From necessity I have been obliged to study economy of labour and for some years have used a hive of the simplest possible construction. Similar in appearance and construction to those used in the U.S.A. and Canada, it consists of a simple floor on which are placed square brood boxes and supers, all of the same horizontal dimensions, surmounted by a framed glass quilt (not essential) and a flat-topped cover. Excluders and super-clearers are of the same dimensions, so that everything fits on everything else. There are no outer cases, plinths, porches, or ornamental parts. Brood boxes are of $\frac{5}{8}''$ timber, made to hold 11 British standard frames, and are $17\frac{3}{4}''$ square. Most of the hive manufacturers now sell similar hives. The "Simplicity" and "National" single-walled hives are good examples. I propose to describe my method of swarm prevention as applied to such a hive and afterwards to show how it may be used, with some modifications, on any type of bar-frame hive.

As soon as the spring has well opened and the stock shows signs of normal development it is stimulated by slow feeding until almost the whole of the brood-box is occupied by bees. It is then supered over an excluder in the ordinary way so that advantage may be taken of an early honey flow should one occur. (Plate I, Fig. 1.) The bees may easily be kept from swarming until the last week in May by giving the queen additional laying room in advance of requirements. This is done by the addition of one or more brood boxes fitted with drawn combs or foundation. An ordinary stock in an average season will need one additional brood chamber. (Plate I, Fig. 2.) A strong stock in a good season may need two or three. The aim must be to get as powerful a colony as possible before

PLATE I

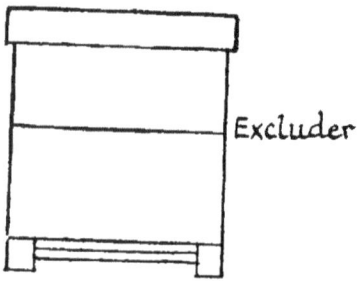

Fig 1.

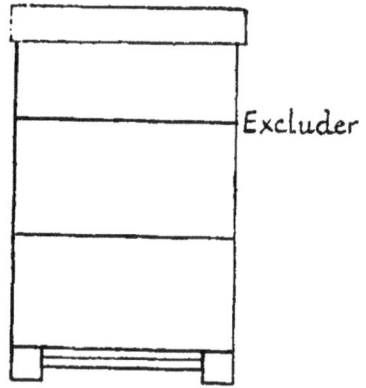

Fig 2

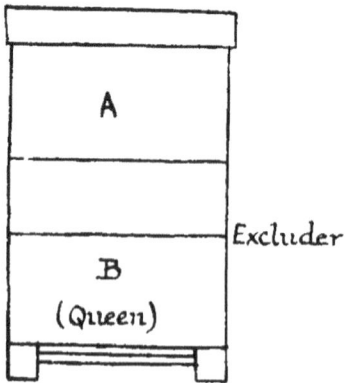

Fig 3.

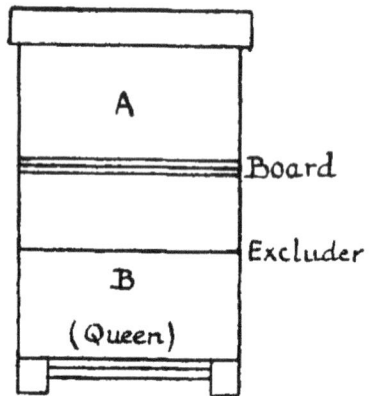

Fig 4.

the last week in May and to do this without causing it to be so crowded that it develops the swarming impulse.

When a new brood chamber is added some precautions are necessary. It is often useless merely to place it *under* the brood nest, for in unfavourable weather the bees will confine themselves to their original combs, prepare queen cells, and swarm. The new brood chamber should be placed *above* the original one, and to ensure that the bees will occupy it two frames of brood with bees should be lifted from the lower to the upper chamber and placed on either side of a central empty comb, their places below being filled by two empty combs or foundation. The same procedure should be followed if it becomes necessary to add further additional brood boxes.

A stock occupying fairly fully a double brood chamber, and possibly to some extent a super, at the beginning of the last week in May is in ideal condition for successful treatment. Better results will naturally be obtained from a stock which has needed a third brood box. During the years 1931–2 I found two brood boxes sufficient, but in May 1933 the weather was so favourable that the bees developed with startling rapidity and boxes were added at a disconcerting rate. It is in matters like this that the beekeeper must adapt his procedure to circumstances and not rely on dates or precise instructions. (Vide pp. 33–34.)

METHOD I

Let us now assume that towards the end of May our typical stock occupies twenty brood frames and that above these there is a shallow frame or section rack, the hive having been well ventilated below by a wide entrance and the queen not having been at any time restricted for room to lay. (Plate I, Fig. 2.) Such a stock will not have prepared to swarm but if left alone will very soon do so. We therefore proceed to prevent this in the following way:—

If possible, for greater convenience in working, provide two spare empty brood boxes and place them on improvised stands (e.g. flat hive covers). Call these boxes A and B.

OPERATION I.

Separate the combs of the double brood chamber into two sets. Place in box A the combs (with bees) containing brood, and in box B the remaining combs (with bees) which do not contain brood. See that the queen, and also a comb containing a little unsealed brood, are placed in the centre of B. Now rebuild the hive, putting box B on the floorboard, an excluder on B, the super above the excluder, and box A above the super. (Plate I, Fig. 3.)

This operation is the main part of the method and usually takes about 15 minutes to perform. It can be effected without the two extra brood boxes, but not so easily. No bees are shaken from the combs. The chief difficulty is to find the queen. To do this the following procedure is helpful:—

Remove the super and excluder and cover frames with a quilt. Smoke the bees gently at the hive entrance two or three times at intervals of a minute. This will cause the queen to go into the top box (which may be temporarily removed) where she will probably be found on the frame of youngest brood. Put her and the frame temporarily into a nucleus hive or at once into box B and then quickly separate all the remaining combs.

If there are more than ten frames of brood for box A the surplus ones are used to strengthen a weaker stock, the poorest being selected for this purpose. Box B can then be completed by some empty combs or frames of foundation. It is essential that box B contain no brood except the little in one frame which is inserted with the queen, but it should contain some stores.

As box A is to be used for queen rearing the arrangement of the brood combs is of some importance. Those containing the youngest brood should be placed towards the middle and the sealed ones towards the outsides. As a rule the brood frames will be well provided with honey and pollen,—both essential for the raising of queen cells. If honey is not present in sufficient quantity a syrup feeder should be placed over the box for reasons which will appear later.

Great care must be taken, while the combs are being separated, to see that the stock has not started queen cells in preparation for swarming. If any should be found the procedure should be as described for Method II, p. 40.

Box B is now in position for the rest of the season. It is to become the new brood chamber of the stock. The small batch of young brood keeps the queen contented and she continues her laying without interruption.

The hive is now left for three days during which interval practically all the nurse bees pass up through the excluder and super to the brood above where they begin to raise queen cells. The queen remains below and the flying bees proceed to the fields as usual.

The flying bees are now to be separated from the remainder of the stock. This is quickly done by means of a specially devised screen board.

THE SCREEN-BOARD.

The screen-board (Plate II) is simple and easily made. It consists of a sheet of three-ply wood of the same dimensions as the top of a brood box. Around its edge and on both sides of it are nailed battens $1\frac{1}{4}'' \times \frac{3}{8}''$. A hole is cut in the middle of the sheet of plywood of

PLATE II

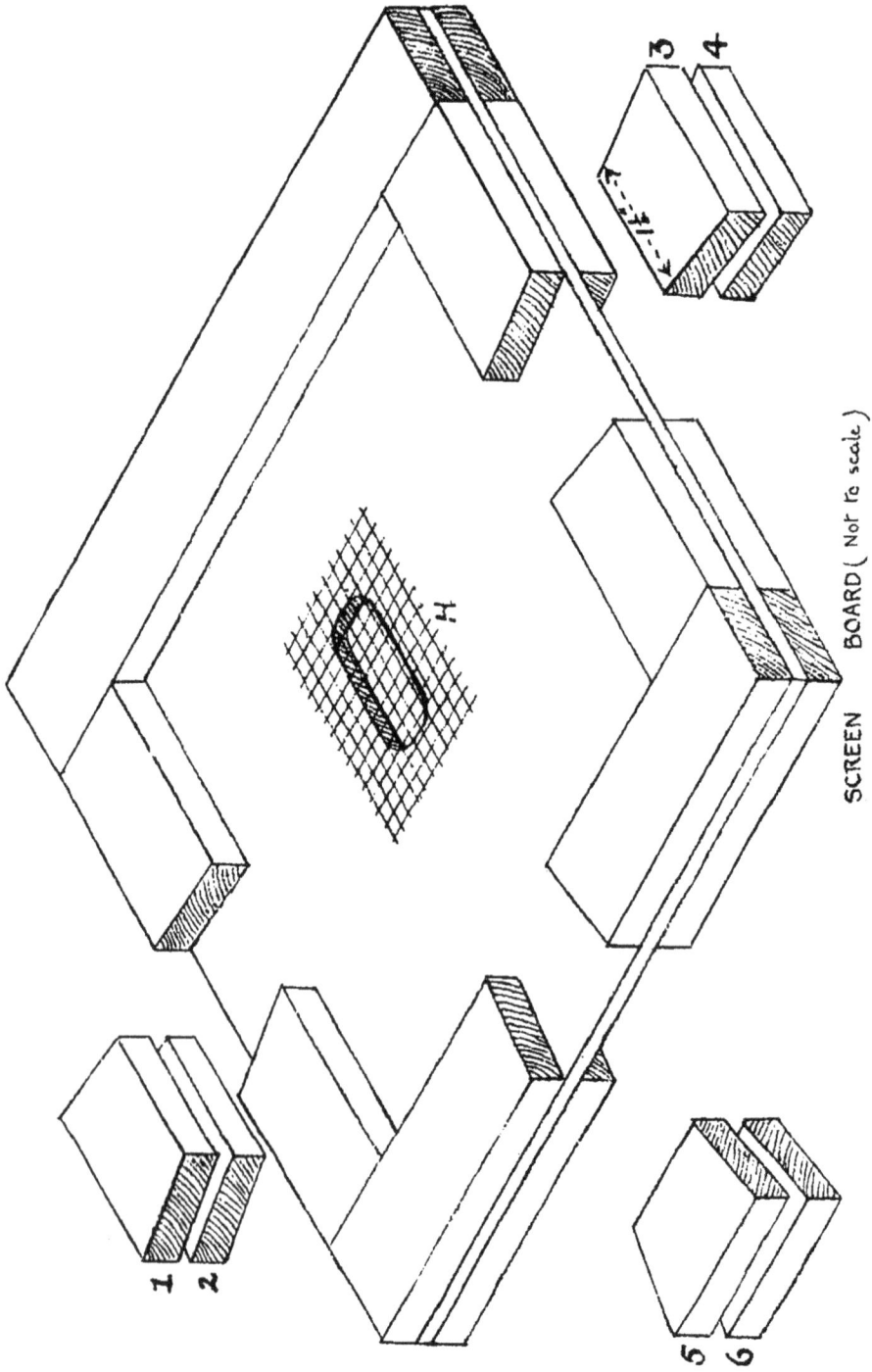

SCREEN BOARD (Not to scale)

the size to seat an ordinary Porter bee-escape. The reader will perceive that this description applies to an ordinary super-clearer which may be adapted for our present purpose without affecting its use as a super-clearer. Conversely a specially made screen-board may be used as a super-clearer. If it be not desired at any time to utilise the screen-board as a super-clearer a larger hole (e.g., circular, of $3\frac{1}{2}''$ diameter or even larger) may be cut at the centre.

On each of three edges of the screen-board rectangular or wedge-shaped pieces, about $1\frac{1}{4}''$ wide, are cut from the middles of the upper and lower battens. On the fourth side the battens are not cut. There will thus be three openings above the plywood diaphragm and three below it when the wedge-shaped pieces are removed. These openings are to serve as entrances for the bees. They can be opened or closed by the removal or insertion of the wedges. If desired the latter may be provided with small screw eyes so that they may be tied by string to the board in order that they may not be lost. They may be Vaselined to make them easily removable. A fourth pair of openings provided in the front edge of the board may prove useful at certain times. (See pp. 31, 35, and 98.)

No bee escape is used but the hole H is covered by a piece of wire cloth or perforated zinc fastened to the board by drawing pins.

OPERATION II.

Three days after the first operation place the screen-board under box A, that is, between it and the super. Place it so that the edge with no small entrances is towards the front of the hive. (Plate I, Fig. 4.) Withdraw No. 1 wedge, leaving all the others in their positions in the board.

This operation can be performed in two or three minutes without disturbance to the bees. Now let us consider what will happen. The bees in A are shut off from those in B. The field bees that happen to be in A however can fly from the small entrance 1. Within a day or so they will all have left A and returned by the main hive entrance into B where they will join the other field bees and the queen. Only young bees will remain in A, whilst all the field bees will be in B and the super.

We shall therefore have a strong artificial swarm in B. The queen, encouraged by the presence of a little unsealed brood, continues to lay as though nothing had happened. Owing to the presence of nurse bees the development of a new brood nest proceeds normally. If honey is coming in the bees store it freely, not needing much for the preparation of brood food. It will be three or four weeks before B is filled with brood and at least five weeks before there is a considerable force of nurse bees. Before this time however the queen will have declined in her laying powers and the stock will not then be likely to develop the swarming impulse.

In box A the bees, although actually in contact with the stock below through the wire cloth in the screen-board, consider themselves queenless, and complete the queen cells already begun. For this work they are under favourable conditions, being young and well provided with honey and pollen. The wire cloth serves two important purposes:—

(1) It allows heat to rise to A from the stock below so that the temperature throughout the hive is fairly uniform and unsealed brood in A cannot be chilled;

(2) From the moment the screen-board is installed a column of bees occupies the super and constantly maintains contact, through the wire cloth, between the queen and stock in B and

the brood and bees in A. If the stock is strong the whole super is occupied at once. This is one of the most interesting and valuable features of the method for it ensures the occupation of the supers whatever these contain,—combs, foundation, starters, or sections. Bait combs are unnecessary, and the chief difficulty of the comb-honey producer, that of getting his bees to take to sections, is overcome. Not only is the super occupied, but experience has shown that if honey is coming in it is promptly stored there.

If there is a honey flow a strong stock may need a second super at the first operation and others must be added as necessary. In the case of a weaker stock it is well to crowd the bees into the super by giving less room in B. If the beekeeper does not mind an extra operation it is advisable on the first day to provide even a strong colony with six combs only in B, placing these in the middle of the box, filling the spaces with dummies and giving necessary super room. Subsequently four other combs may be added as considered necessary,—say two in a fortnight's time and two others a week later. This will ensure that, from the beginning, incoming nectar will go to the supers.

Now let us consider what happens in A. The young bees, deprived of their queen first by the excluder and subsequently by the screen-board behave as though queenless, and at once begin to build queen cells. The warmth from the bees below the board is favourable, and although the field bees have gone, young ones are "hatching" daily in great numbers.[1]

In a few days' time a considerable force of field bees will have developed in A. It is desirable that these should join the stock below and take their part in

[1] Vide Appendix, p. 107.

filling the supers. On or about the seventh day therefore the following simple operation is performed:—

OPERATION III. (7th or 8th day)

Replace wedge 1 and remove wedge 2. Remove wedge 3 on the opposite side of the hive.

The field bees from box A must now leave by opening 3 but when they return they attempt to enter the hive at opening 1 as they have been accustomed to do. Finding this closed they enter at opening 2, which leads to the lower stock in B. Henceforth this opening will be their means of exit and return. The stock in B thus receives an important reinforcement, and for the time being there are no field bees in A.

At this stage it is as well to look at some of the combs in A to make sure that the bees are building queen cells. If these are inferior this may be due to bad weather, lack of income or of nurse bees. In this case and if superior young queens are desired, Operation III should be omitted, all the queen cells in A broken down and a fresh comb of eggs or very young larvae inserted, on which better queen cells will be raised (p. 29).

A week later a further force of field bees will be seen working from opening 3. These also are added to the lower stock as follows:—

OPERATION IV. (14th or preferably the 15th day)

Replace wedge 3 and remove wedge 4. Remove wedge 5 at the back of the hive.

Field bees accustomed to use entrance 3 now leave at 5, and on attempting to return by 3 enter at 4, which leads to the lower stock. In this way the latter is again reinforced, and again there are no field bees in A. Now at this time, usually on the 14th day, the young queens will be emerging from their cells. Ordinarily this would occasion one or more small swarms, but as the colony has been deprived of its flying bees it cannot possibly swarm, and finally, after the usual destruction of rivals, one young queen heads the colony. She will fly to be mated after she is four days old, finding exit by opening 5. A small coloured alighting board may with advantage be temporarily fixed under this entrance.

It should be specially noted here that it is not necessary for the beekeeper to destroy queen cells. He leaves that to the bees.

Formerly I arranged for opening 5 to be in the front of the hive and I imagined that some young queens lost on mating flights had entered the main entrances of the lower stocks, attracted thither by the greater crowd of flying bees. There they probably deposed the old queens and reigned in their stead, leaving the top stocks queenless. With opening 5 at the rear of the hive this risk is eliminated.

In the first edition of this book I recommended that Operation II be carried out on the 2nd day. Experiments carried out by Capt. Harrison and myself have shown that better queen cells are made in box A if the screen-board is inserted on the 3rd or 4th day. The busy beekeeper may, if he likes, omit Operation III; that is to say he may postpone it to take the place of Operation IV on the 14th or 15th day (p. 28).

If bad weather prevents the flying bees from leaving A on the 15th and subsequent days tiny swarms may issue with virgin queens. This may be prevented by gently shaking a few bees from some of the combs in A before the main entrance to B. If a comb is shaken sufficiently gently the older bees fall off, leaving the young bees adhering to them. The combs containing queen cells should not be shaken.

A spell of cold weather or a lack of food may deter the young bees in A from raising queen cells. In such a case they should be provided with a fresh comb containing eggs or very young larvae (p. 28).

In the event of the loss of the young queen in A the

bees may be given a little young brood from which to raise another, or alternatively, about five weeks after the first operation, they may be shaken down in front of and united to the stock in B.

It is important to bear in mind that the stock in A, being deprived of its field bees, cannot store nectar for some time and therefore must be fed if necessary.

In 1933 I divided each top box A into three nuclei by means of close fitting division boards and separate entrances. (Vide Chap. III. 7.) Every hive provided at least one fertile queen, many gave two, and some gave three.

Should a stock swarm before treatment it should be hived on empty combs under the screen-board. The parent colony should be placed above the board, with one exit, and the second reinforcement made seven days later. The next reinforcement, if any, should not be effected until after the young queen in A is laying. This plan gives excellent results.

The method provides many possibilities in queen rearing which may be left to the imagination of the reader, but one is of special value. If several hives are being dealt with it is easy to requeen from the best colonies. These should be treated two or three days earlier than the others and on the 12th day afterwards their spare queen cells should be carefully cut out and one of them inserted between the frames of each other colony whose queen cells would be in a less advanced stage. The first queen to emerge would head the colony.[1]

Let us consider again for a moment the condition of the main stock in B. It is really a strong artificial swarm, headed by the old queen, which has twice been heavily reinforced by the flying bees from A. In respect of field bees it is as populous as if the original stock had remained undisturbed and had not attempted to swarm. Occupying a limited brood nest at first its honey goes

[1]Vide Appendix, p. 107.

to the supers. Beginning with few young bees and never having an excess of them it does not swarm. Its strength has increased in contrast with that of a natural swarm which gradually diminishes during the first month. Care must be taken, as in the case of a natural swarm, that the stock is not without food should the weather become unfavourable. To feed B, give stored combs; or remove A and screen-board to a temporary stand. Feed B rapidly with warm thick syrup and then restore screen-board and A to their position over the supers.

It will be noted that as there are two "top entrances" in openings 2 and 4, thorough ventilation, which is considered to be a deterrent of swarming, is assured. As the season advances these small entrances become less used, and if sections are being worked they may then with advantage be closed. The bees returning to them readily find their way to the main entrance.

When the young queen has become mated and a good brood nest has developed in A it may be desirable again to reinforce the stock below in order to take advantage of a late honey flow. Entrance 6 may be used for this purpose. As it will then be necessary to use entrance 1 or 3 over again the corresponding lower entrance 2 or 4 must be closed a few days previously.

If a super-clearer be used to remove filled supers while the screen-board is in use the side entrances must first be closed.

Wedges may be extended so as to project beyond the board for easier handling. They may also be made with a hexagonal section to prevent them from falling out when the board is removed.

Many beekeepers would desire to perform Operation I without having the trouble of searching for the queen. This can easily be done in the following manner:—

Separate brood combs from the others as described

on page 22 without looking for the queen. Place excluder and super in position on box B. Shake the bees off the brood combs in front of the main entrance. If the queen is there she will run into B with the bees. If not, she is in B already. Place A containing the brood combs over the super as previously directed. The nurse bees will pass through the excluder to occupy these combs before the next day, leaving the queen below.

Brood combs must be shaken gently or the larvae will be dislodged from their proper positions. It is not even necessary to shake all the bees from the combs. Many young bees will still adhere to them and may be left there. The queen almost invariably falls at the first ".shake". If the combs contain freshly gathered nectar they should not be shaken or the bees may be drowned in the falling nectar. In this case it is much better to find the queen.

Concise directions for carrying out the method are given on pp. 38–39.

The main essentials for success in applying the above-described method are:—

(1) It must be used *before* the bees develop queen cells in preparations for swarming.

(2) The bees must not go short of food in a bad season.

(3) Ample super room must be provided.

As the actual work involved in the whole method is small it is well to perform Operation I on as many hives as possible in one day. At the same time the dates for Operations II, III and IV should be recorded on each hive to avoid subsequent confusion. If there are many hives it is well to have the help of an assistant for Operation I. For subsequent operations the work and time needed are trifling and no help is needed.

The honey season ended supers are removed and the beekeeper decides whether he desires increase or not. The stocks in the boxes A will have become strong enough to be removed to separate stands. If however increase is not desired stocks A and B can be united by removing the piece of wire cloth in the screen board. The bees unite peaceably and the old queen if not removed by the beekeeper may be assumed to be destroyed by the young one. Later, when cool weather has confined the bees to the hive for some days the screen board should be removed and the stock reduced to one brood chamber for the winter.

In some cases the bees are not sufficiently strong at the end of May and then the application of the method is deferred till later. If very late the capacity of the new brood nest in B should be correspondingly limited.

In the case of a strong stock which has needed three or more brood boxes before the application of the method the procedure under Operation I is modified as follows:—

Separate the combs into three lots and place

(1) The combs of youngest brood in box A;
(2) Broodless combs, queen, and one comb with a little unsealed brood in B;
(3) Remaining combs, empty, containing honey, or with sealed brood only, in an additional box to be used as a super. Care must be taken that there are no larvae or eggs in this box or a young queen will be reared in it, become mated, and fill the supers with brood.

Alternatively the whole of the remainder of the stock, even if it includes two or three boxes, after the arrangement of box B, may be placed above the screen-board, the youngest brood being placed in the lowest of the boxes A. As the upper boxes of A become broodless they may be removed and used as supers or otherwise. This plan is easy and ensures ample stores in A.

A strong stock headed by an unusually prolific queen may need some relief of its brood nest three weeks after the first operation. This may be effected by exchanging some frames of sealed brood in B for empty ones in A. Similar relief will be needed if a stock is operated upon too early in the season.

STOCKS FOR THE HEATHER.

Stocks managed by the foregoing method may be made exceedingly strong by the union of the stocks in A and B before being taken to the heather.

The brood nest B can be completely filled with brood from both boxes and all the bees in A can be shaken down into B. This should be done on the evening before the stock is to be removed so that the bees may be confined at once. The united stock will need sufficient supers and adequate ventilation.

In some years, such as 1935 and 1936, continued wet weather in May and June may cause bees to develop the swarming impulse in July or even in August. Even in normal seasons a stock headed by a prolific queen may throw a late swarm after having been prevented from swarming during the usual period.

To prevent such late swarming it is a good plan to get the young queen below as soon as possible. This can be done as follows:—

Let us assume that Method I has been successfully applied and that the young queen in A has become fertile and developed a fairly large brood nest.

On a fine day when the bees are flying freely *and bringing in nectar*, exchange the positions of boxes A and B as they are. (There may be some fighting if this is done on a day when no nectar is coming in.) Signs of incoming nectar are:—

(1) Returning field bees drop heavily on, and crawl slowly up the alighting board.

(2) At night a roar of fanning bees is heard and a strong odour emanates from the hive entrance.

The flying bees will now join the young queen below, and however much they may become crowded, as when working sections, there will be no likelihood of late swarming.

(To ensure greater security for the young queen the exchange of boxes A and B may be made late in the evening when flying has practically ceased, a sheet of newspaper pierced in a few places by a small nail or match-stick being placed under the excluder.)

Mr. R. E. Kisby, of Thetford, Ely, who, amongst many others, has successfully used this plan, recommends the use of a fourth pair of entrances in the front of the screen-board—Nos. 7 and 8. At the time of exchanging the boxes he closes 5 and opens 6 and 7, thus securing for the stock below all the bees flying from the back of the hive.

At any time subsequently, if desired, the old queen may be removed and her stock added as a super to the stock below.

ALTERNATIVE TO METHOD I

A modification of this method, based on the fact that a young queen rarely swarms in her first season from the stock in which she was bred, was suggested to me by Mr. I. Whiting, of Rockford, Illinois, U.S.A. It involves the use of my screen-board in conjunction with a "supersedure" method described in *Gleanings* some years ago by Mr. C. A. Wood, of South Wayne, Wisconsin, U.S.A. It worked perfectly in the case of several stocks in the almost honeyless season of 1936, and I have since used it with good success. It may be recommended to those who would avoid the necessity of finding the queen and keeping to fixed dates.

The procedure is as follows:—

On a day when the bees are flying freely take one or two combs of very young brood and eggs (without bees) and place them in an empty brood box. Fill this box with broodless combs or foundation. Two or three of the combs at least should contain honey and pollen. Place this box over an excluder above the stock until nurse bees have ascended to occupy the two brood combs. A few minutes are usually sufficient for this.

This box is to be the future brood nest of the stock. Place it on the floorboard, and above it, in succession, excluder, supers, screen-board, and the remainder of the stock, including the queen. Open the upper entrance at the *back* of the screen-board. All the flying bees will join the new stock below, which will rear a young queen, and may be re-inforced at convenient intervals from the stock above by means of the other entrances in the screen-board.

Sometimes, as in the case of Method II (see p. 42), the flying bees discover that their queen is above the board and may even travel from the front entrance round the hive to find her. They can be made to leave, however, by removing the upper stock to a new position for two days, after which it may be restored to its proper position.

From seven to ten days later reduce queen cells to one, or give a ripe cell from a selected colony.

This method must not be used, of course, if the stock has begun preparations for swarming.

PLATE III

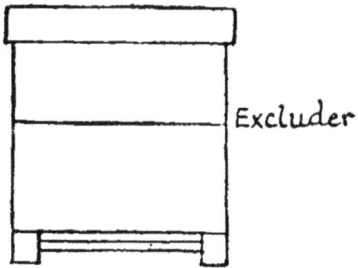

Fig 1.

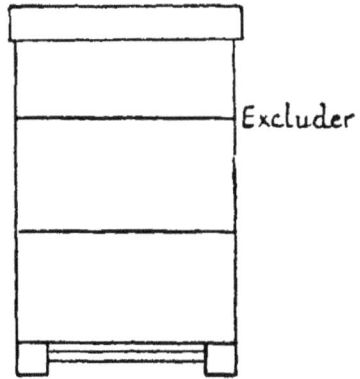

Fig 2

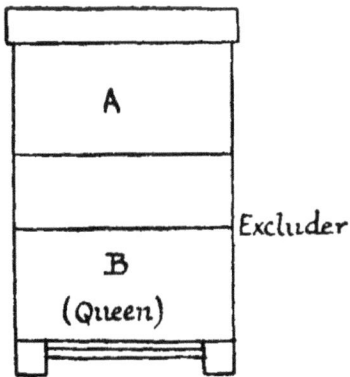

Fig 3.

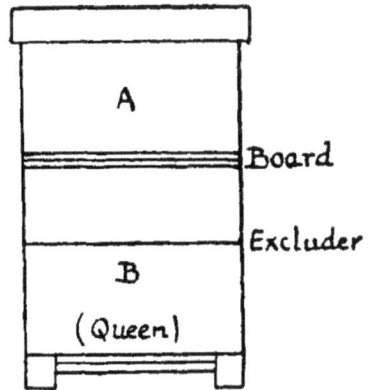

Fig 4.

SUMMARY OF METHOD I

For a stock in any single-walled hive,—*without* queen cells.

1ST DAY.

Separate combs and place them as follows :—

In box A :—

Combs containing the brood, with adhering bees.

In box B :—

(a) Broodless combs, with adhering bees.

(b) One comb with a little unsealed brood.

(c) The queen.

Rebuild the hive in the following order :—

Floorboard, Box B, Excluder, Super, Box A. (Plate III, Fig. 3.)

4TH DAY.

Place screen-board under box A, with its edge without small entrances towards the front of the hive. (Plate III, Fig. 4.) Withdraw wedge 1, leaving others in position in the board.

7TH OR 8TH DAY.

Replace wedge 1 and remove wedge 2. Remove wedge 3 on the opposite side of the hive. Ascertain if queen cells are being built in A (see p. 28).

14TH OR 15TH DAY.

Replace wedge 3 and remove wedge 4. Remove wedge 5 at the back of the hive.

METHOD II

(Applicable to stocks which have started queen cells)

As a rule the extension of the brood nest is sufficient to prevent the development of the swarming impulse until the end of May. Occasionally, however, the bee-keeper, on examining a hive, finds that notwithstanding his precautions the bees have queen cells in a more or less advanced state, and he realises that if left to themselves they will swarm as soon as the ripest queen cells are sealed. I venture to say that in these circumstances few, if any, beekeepers know how to prevent a swarm without breaking down the queen cells and either caging the queen or taking her away, and by so doing injuring the stock for a succeeding honey flow. The method described in the preceding pages has some-times proved effective, the queen being left below in box B without brood, and all the queen cells being placed in box A with the remainder of the brood. In most cases, however, the bees in B have continued to raise cells as soon as young brood has become available. Even further extractions of nurse bees from B have not always proved effective.

At the beginning of June, 1933, four of my hives had developed queen cells before I could deal with them. I applied Method I to them, leaving no brood with the queen in the bottom box, hoping this would work satisfactorily as it had done in the previous year.

On examining the top boxes of two of these hives a few days later I found that the queens had forced their way through the queen excluders and had joined the young bees in the top box where the queen cells and brood had been placed. To my surprise the queens were laying well, and what was far more important, the queen cells were being nibbled down.

Had I found a simple way of dealing with a stock which had prepared queen cells?

I at once proceeded to the other two hives and *put the queens up with the queen cells.* The result was the same, for in both cases the queen cells were all destroyed within the ensuing week.

Subsequently two other stocks developed queen cells before they could be dealt with. Method I was applied to them but the queens were put into the top boxes with the queen cells and again the latter were all broken down by the bees.

I had now found a simple modification of the method which would enable me to apply it whether the bees had previously prepared queen cells or not.

In the case of stocks which have developed the swarming impulse therefore the procedure should be as follows:—

Proceed as advised for Operation I (page 22), but place all brood and queen cells, as well as the queen, in box A. Into box B put the broodless combs, with bees, and also one comb of sealed brood in the centre. It will not matter, and may even be an advantage, if this central comb contains a few mature unsealed larvae, but great care must be taken that no eggs or young larvae are present. When rebuilding the hive place the excluder immediately under A to keep the queen out of the super. (Plate IV, Fig. 5.) If any of the queen cells are sealed there is no time to lose and the screen-board must be placed in position (Operation II) at once, the sealed cells only destroyed, and the excluder placed in its proper position above B. (Plate IV, Fig. 6.) If they are all unsealed this should be done on the second day.

The field bees will now leave box A and rejoin the queenless bees in box B below where they will quickly lose the desire to swarm. Aware of the presence of the queen above the board they will not become disheartened but will continue their work in the fields. Meanwhile the queen and young bees in A, finding themselves deprived of flying bees, do not swarm and gradually destroy the queen cells. To hasten this, *Operation III is effected five days instead of seven days later.* When the queen cells are all destroyed and the queen is again laying normally,—usually before the end of a week,—*she and the comb with bees on which she is found should be transferred to box B.* (Plate IV, Fig. 7.) She will be readily accepted there without being caged and the swarming impulse will have disappeared.

The young bees in A will now build a new lot of queen cells on eggs recently deposited by the queen. *Operation IV should be applied fourteen days after the transfer of the queen.*

The important point in this method is to get the flying bees to quit box A. Occasionally they discover where the queen is and some of them return to her. They can be made to leave completely by moving the box to a position some yards distant for two days, placing it preferably near another (untreated) hive to avoid the loss of a few flying bees when the box is restored to its proper position above the screen-board. Such removal may be effected on the 1st or any succeeding day, but certainly on the 5th or 6th day if by that time it is found that the queen cells are not being broken down. Usually there is no difficulty and the flying bees readily quit box A via the screen-board, especially if the exit at the back is used first. (See order of using wedges, p. 45.) Should bad weather prevent the bees from leaving A some should be shaken into B. (See p. 29.) This may well be done at once if queen cells in A are sealed. But in this case it is preferable to use Method IV (p. 61).

If eggs or young larvae have been overlooked and inadvertently left in box B queen cells will be built there. These should be removed when the queen is transferred to the box on or about the 7th day.

For convenience the method is summarised on page 45.

PLATE IV

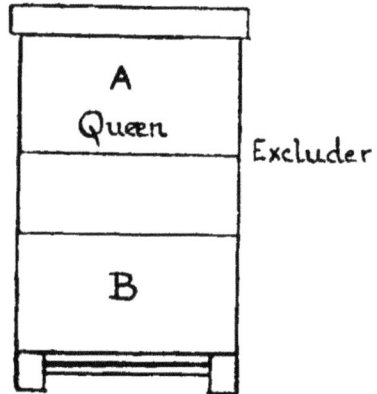

Fig 5.

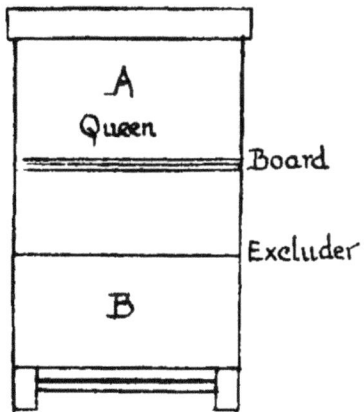

Fig 6.

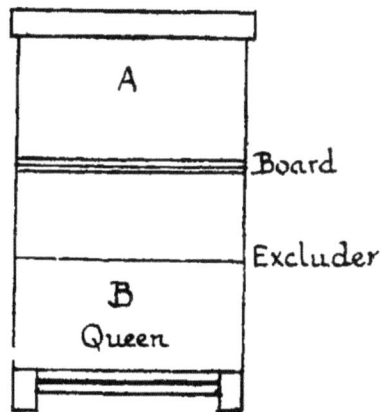

Fig 7.

SUMMARY OF METHOD II

For a stock in any single-walled hive,—*with* queen cells.

1ST DAY.

Separate combs and place them as follows :—

In box A :—

(*a*) *The queen.*
(*b*) *All the brood, with adhering bees.*
(*c*) *All the queen cells. (Vide Method IV, p. 61.)*

In box B :—

(*a*) *The combs without brood, with adhering bees.*

(*b*) *One comb of sealed brood.*

Rebuild the hive in the following order :—

Floorboard, Box B, Super, Excluder, Box A. (Plate IV, Fig. 5.)

2ND DAY (or on 1st day if there are *sealed* queen cells).

Place screen-board under A and the excluder between B and the Super. (Plate IV, Fig. 6.) Withdraw wedge 5 leaving the other wedges in position in the board.

5TH DAY.

Replace wedge 5 and remove wedge 6. Remove wedge 3 on the side of the hive.

7TH DAY (or when queen cells have been completely destroyed).

Transfer the queen with a frame of brood and bees from A to B. (Plate IV, Fig. 7.)

14TH DAY AFTER TRANSFER OF QUEEN.

Replace wedge 3 and remove wedge 4. Open wedge 1 at the other side of the hive.

METHOD I (A)

Modification of Method I, applicable to any hives, particularly double-walled hives, for which it is not desired to make a screen-board, and in which queen cells have not been started.

Methods I and II described in the preceding pages can be applied to all single walled hives provided that brood boxes, excluders, supers, and clearer-boards all have the same horizontal dimensions.

They can be applied also to W.B.C. (double-walled) hives if suitable screen-boards are contrived for the purpose. Mr. F. Gladwell, of Downend, Bristol, makes a screen-board of three sheets of plywood separated by battens which provide tunnels extending to apertures in the lifts.

Messrs. Burtt & Son, Stroud Road, Gloucester, make a simpler board for W.B.C. hives in the adaptation of which cutting or alteration of the lifts is unnecessary.

Any beekeeper who is prepared to provide himself with such screen-boards need not concern himself with the alternative methods which follow.[1]

The principle on which these alternative methods depend is the same as before,—the separation during the swarming period of the young bees from the flying bees. No screen-board is used and therefore the advantages of a specially favourable temperature for the rearing of young queens and the certain attraction of the bees into the supers are not obtained. Swarming however is definitely and easily prevented.

The procedure is as follows:—

The stock is built up during the spring in the usual way.

OPERATION I. (1st day.)

This is carried out as described for Method I on p. 22. (Plate V, Fig. 8.)

[1]Vide Appendix, p. 107.

OPERATION II. (4th day.)

Practically all the nurse bees will now be in box A. *Remove box A and place it on another stand by the side of, and near to the parent hive.* (*Plate V, Fig.* 9.) All flying bees will leave A and enter B where they will find the queen and remaining field bees. The nurse bees left in A will prepare queen cells and the hives may be left alone until the 14th day.

OPERATION III.

On the 14th or 15th day A will have developed a large force of field bees and young queens will be emerging from their cells. *Remove A to a new position, —at least half a dozen yards from the parent stock.* (*Plate V, Fig.* 10.) The field bees from A will join the parent colony and so make it very strong. The young bees in A cannot then swarm. If it is desired to divide A into two or three nuclei so as to increase the chance of young queens being mated this should be done on the 12*th day* and the nuclei removed to new positions at the same time.

Supers should be added to the parent colony as required.

A simple method of dividing A into two nuclei is to insert a closely fitting division board into the middle of the box. See that there are some queen cells in each compartment. Bore a $\frac{5}{8}''$ hole in one side or the back of the hive from which the extra queen may fly to be mated. A first quilt of tough material should be securely fixed by drawing pins to the division board so that the bees cannot pass from one compartment to the other.

An inverted box of convenient size will make a suitable floor for this temporary hive and a piece of tarred felt will serve as a cover.

For convenience the whole method is summarised on page 49.

PLATE V

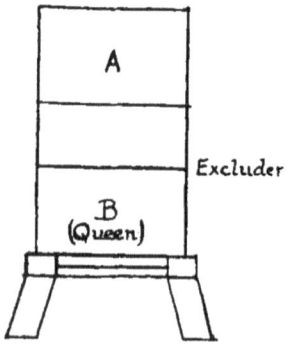

Fig 8.

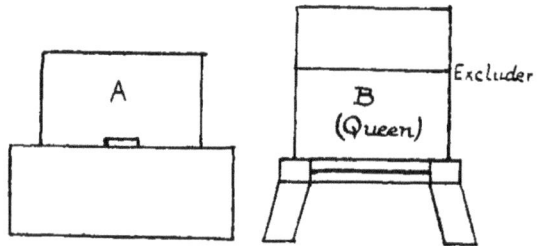

Fig 9.

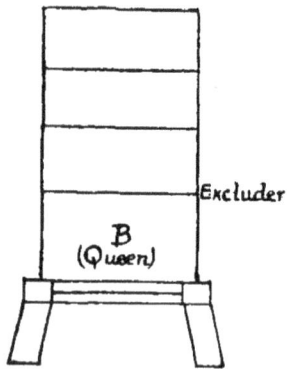

Fig 10.

SUMMARY OF METHOD I (A)

For a stock in a double-walled or other hive,—*without* queen cells.

1ST DAY.

Separate combs and place them as follows :—

In box A :—

All the brood, with adhering bees.

In box B :—

(a) Combs without brood, with adhering bees.
(b) One comb containing a little unsealed brood.
(c) The queen.

Rebuild the hive in the following order :—

Floorboard, Box B, Excluder, Super, Box A. (Plate V, Fig. 8.)

4TH DAY.

Place box A on a new stand by the side of box B (Plate V, Fig. 9.)

14TH DAY.

Remove box A to a new position some yards distant. Add Supers above B as necessary. (Plate V, Fig. 10.) or alternatively:—

12TH DAY.

Divide A into nuclei and remove these to distant positions.

ALTERNATIVE TO METHOD I (A)

The following plan, involving the use of an interior screen-board, and therefore no disturbance of lifts, has been tested during the past two years and has proved generally successful. It was first suggested to me by Mr. S. B. Stokes of Solihull, Warwickshire, and the same principle of working occurred also to Mr. Wm. Shepherd, of Newton Mearns, Glasgow.

The procedure is as follows:—

The stock is prepared as described for Operation I, (p. 22) and a screen-board inserted, as directed, under A. This screen-board differs from that described on p. 23. It should be of the same size as the top of an inner brood chamber, and should be provided with only one upper entrance at the front, 3 inches wide. All the inner boxes are pushed back as far as possible in the hive which is darkened above by closing the bee-escape in the roof.

As the flying bees leave A they descend towards the light and leave the hive by the main entrance, and on their return they join the lower stock in B. If the hive is left undisturbed box A will have no bees left in it about four weeks later and the stock in B, containing the whole population, will have become very strong. Swarming will have been prevented because the stock in B will provide plenty of work for the nurse bees.

The crowded condition of B, however, is likely to lead to late swarming unless steps are taken to prevent it. This can be done by either:—

(*a*)　Repeating the operation three weeks later.

(*b*)　Exchanging combs of sealed brood in B for broodless ones in A two or three times at fortnightly intervals. The number of combs to be exchanged will depend on the strength of the remaining population in A.

If a young queen be reared in A she will take her mating flight from the main entrance, and in almost every case will be balled and destroyed when she enters B. Mr. W. Ireland, of St. Bees, Cumberland, reports one case in which the young queen entered B and caused the old queen to come out with a small swarm. The young queen subsequently headed the colony.

As a rule the issue of a young queen from the main entrance is useless and should be prevented by the destruction of queen cells in A, but if the queen cells are satisfactory requeening can be provided for by means of a nucleus, either external, or partitioned off at the back of A with a flight hole through the lift. This flight hole should be given about fourteen days after the main operation and the depletion of the bees in A should then be stopped by closing the front entrance in the board until the young queen is fertile and laying well. Boxes A and B may then be exchanged (see p. 34) and the bees from the old queen's stock caused to descend to join the young queen below.

Mr. Stokes recommends that A be allowed to raise a queen to be mated from a flight hole through the lift in the front. This facilitates subsequent uniting as no bees will be flying from the back of the hive.

The stock in box A, having no returning field bees, is likely to run short of stores and should be fed if necessary.

By this method a colony is only temporarily divided to prevent swarming, and ultimately becomes exceedingly strong, with correspondingly good results. In the disastrous season of 1936, Mr. H. L. Burtt, of Midford, Bath, extracted 77 lbs. of honey from a stock so treated, as compared with an average yield of 20 lbs. from his other hives.

METHOD II (A)

Modification of Method II, applicable to any hives, particularly double-walled hives, for which it is not desired to make a screen-board, and in which queen cells have been started. (Alternative, Method IV, p. 61.)

If queen cells are found in the hive when the time has arrived to prevent swarming, proceed as follows:—

OPERATION I.

Separate the combs of the double brood box as follows:—

In box A place all the brood with adhering bees, all the queen cells, and the queen. Into box B put the broodless combs, with bees, and also one comb of sealed brood in the centre. (See page 41.) Rebuild the hive, placing B on the floorboard, the super above B, the excluder above the super, and A above the excluder. It is necessary to put the excluder in this position to keep the queen out of the super. (Plate VI, Fig. 11.)

OPERATION II. (*2nd day, or later on the 1st day if any of the queen cells are sealed.*)

Remove box A to a new stand placed near box B which remains on the original stand. Restore the excluder to its proper place between B and the super. (Plate VI, Fig. 12.)

The object of this operation is to cause the flying bees to leave box A. To ensure this the fronts of the two hives must be made dissimilar in appearance. This can be effected by temporarily placing in front of the entrance to box A a small leafy branch of a tree, or other obstacle, putting the box itself a yard or two from the parent stand and turning it in a slightly different direction. When Operation III is performed on the 5th day the obstacle in front of A should be removed.

These precautions are important as the flying bees will rejoin the queen if they can find her, and every care must be taken to prevent this. To make certain of success it is advisable to move Box A to a distance for two days as directed on p. 42.

All the flying bees of the stock will now enter B. They will be queenless and unable to raise a queen and will quickly lose their desire to swarm. The stock in A having lost all its flying bees will begin to destroy the queen cells. This process will be hastened by a further abstraction of the flying bees,—as follows:—

OPERATION III. (5th day.)

Remove A with its stand to the opposite side of B. (Plate VI, Fig. 13.) The few flying bees from A will then enter B. This further depletion of the stock in A will cause the young bees quickly to complete the breaking down of their queen cells. Meanwhile the queen will continue to lay in a normal way.

OPERATION IV. (About the 7th day, or when all queen cells are destroyed.)

When queen cells in A have all been broken down by the bees, transfer the queen, without caging her, and the frame of brood with bees on which she is found, to B.

B will now contain a strong artificial swarm headed by the queen, the bees of which will have lost all incentive to swarm.

OPERATION V. (14th day after the transfer of the queen to B.)

Remove A and its stand to a distant position, thus causing its field bees to reinforce B. (Plate VI, Fig. 14.)

To increase the chances of obtaining a young fertile queen in A the stock in this box may be divided into nuclei on the twelfth day after the transfer of the queen to B and these nuclei removed to new positions.

This method is summarised on page 55.

PLATE VI

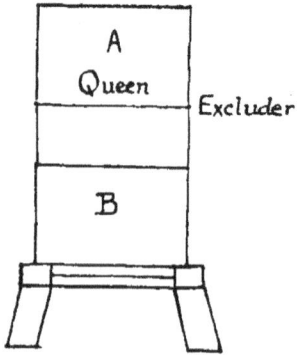

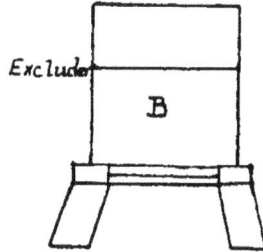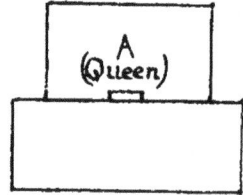

Fig 11.

Fig 12

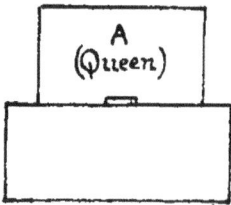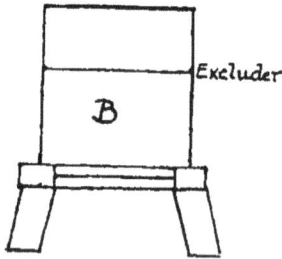

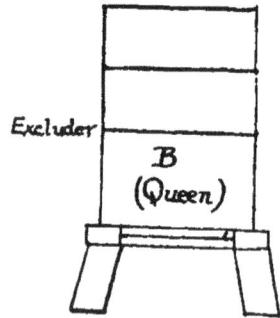

Fig 13

Fig 14.

SUMMARY OF METHOD II (A)

For a stock in a double-walled or other hive,—*with queen cells.*

1ST DAY.

Separate combs and place them as follows :—

In box A :—

(a) *The queen.*

(b) *All the brood, with adhering bees*

(c) *All the queen cells.*

In box B :—

(a) *The combs without brood, with adhering bees.*

(b) *One comb of sealed brood.*

Rebuild the hive in the following order :—

Floorboard, Box B, Super, Excluder, Box A. (Plate VI, Fig. 11.)

2ND DAY.

Remove box A to a new stand on one side of the parent hive. This should be done on the first day if any queen cells are sealed. Place excluder between box B and super. (Plate VI, Fig. 12.)

5TH DAY.

Remove A with its stand to the opposite side of the parent hive. (Plate VI, Fig. 13.)

7TH DAY. (Or when all queen cells have been destroyed.)

Transfer the queen and a frame of brood with bees from A to B.

14TH DAY AFTER TRANSFER OF QUEEN TO B.

Remove A with its stand to a distant position. Add supers to B if necessary. (Plate VI, Fig. 14.)

APPLICATION OF METHODS I AND II
To Modified Dadant and other large hives.

With slight modifications the foregoing methods are applicable to all large bar-frame hives. In these the capacity of the brood chamber may be so great that the bees will not occupy two brood boxes before the queen reaches the peak period of her laying.

For well populated Langstroth hives in a good district no modification is necessary. In less favourable circumstances it may be found desirable to restrict the number of combs in box B by means of dummies so as to force the bees into the supers, and later to provide additional combs if and as needed.

Both Methods I and II have been successfully applied to 12 frame Modified Dadant hives by Capt. A. G. Harrison of Callington, Cornwall. He waits until the bees are working well in at least one super and then applies the appropriate method, limiting the capacity of both boxes A and B to seven combs each. In each case the combs are kept towards the back of the hive, the front portion of B being filled by a large hollow 3-ply dummy (with bee space beneath) and that of A being shut off by a division board which fits down on the screen-board. The youngest brood in A is placed as nearly as possible over the hole in the screen-board.

The supers are placed so that their frames are at right angles to the brood frames. The small portion of the surface of the super which is not covered by brood frames prior to the insertion of the screen-board is temporarily covered by a small quilt or board.

As the conditions of climate and district, and strains of bees vary enormously, the beekeeper must necessarily use his discretion as to the number of combs to be utilised in boxes A and B. He should however limit the capacity of B in the first instance so that incoming honey will be stored in the supers.

CHAPTER II

SWARM CONTROL WITHOUT OPERATION

METHOD III

MANY people wish to avoid manipulative means of swarm control. There is no certain method for this, but if the following suggestions are observed the risk of swarming is reduced to an absolute minimum.

(1) Keep the hive in complete shade.

(2) Provide a queen in her first full year.

(3) Winter in a full-sized brood box plus a shallow food chamber (Pl. VII, Fig. 1, A, B) and allow a full-sized entrance protected from mice in winter and spring.

(4) In spring, about early fruit-blossom time, super with drawn combs, preferably wet from last year's extracting, over an excluder.

(5) As soon as warm days come raise the front of the hive from the floor by 1" blocks at the corners. This allows front and side exits and keeps the hive temperature down.

(6) When the super is well occupied place another —partly of foundation—underneath it. Add other supers as necessary.

(7) If the stock is strong and the queen young, add another shallow worker comb super *to the brood chamber*; that is, under the excluder.

(8) If very warm weather comes raise the hive from the floor by 1" blocks at the remaining corners.

(9) If a good honey flow comes and the stock is strong an additional super of drawn combs may be added, *and at the same time* an additional shallow box of worker foundation added to the brood nest under the excluder. This last operation, which increases both brood chamber and the supers at the same time, finally ensures that there will be no unemployed bees and therefore no swarming.

The last box of combs added to the brood chamber may finally be stored with honey if the season be sufficiently good.

PLATE VII

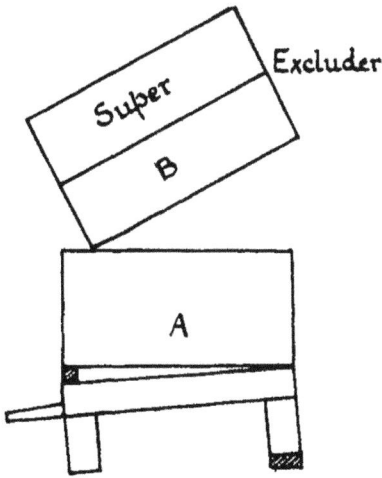

Fig 1.

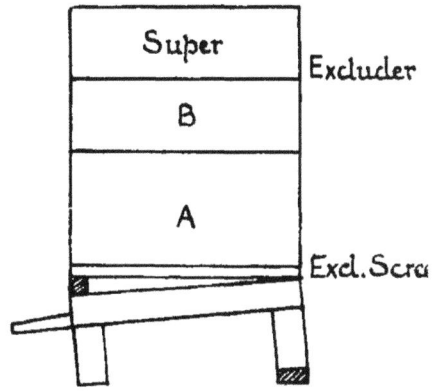

Fig 2

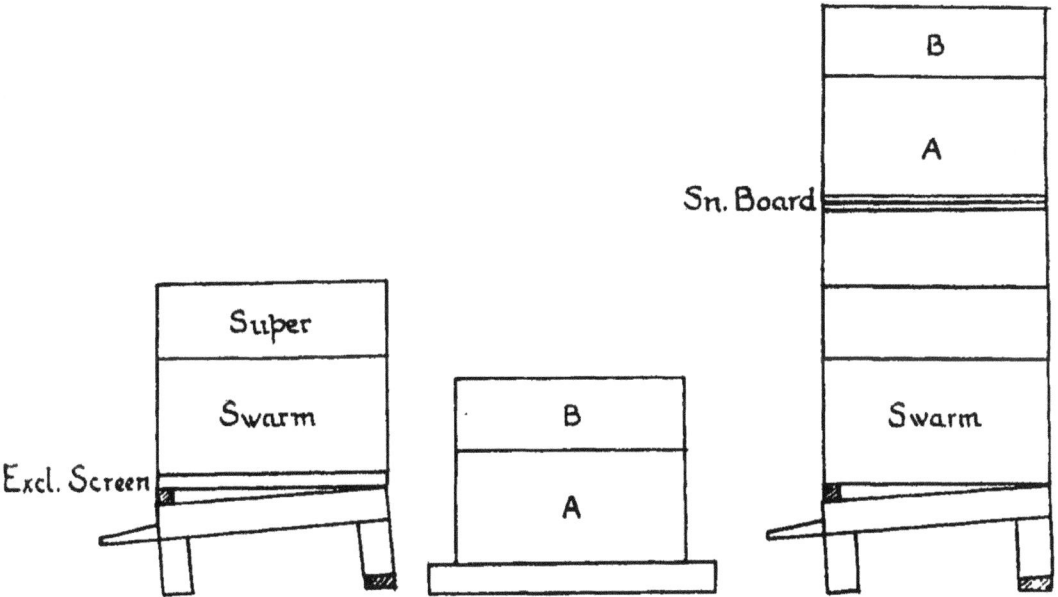

Fig 4   Fig 3   Fig 5

CHAPTER III

CONTROLLED NATURAL SWARMING

METHOD IV

THIS method, demanding little labour and time, and keeping closely to Nature, is specially suitable for swarm control in apiaries of any size, since *only the stocks which prepare for swarming have to be operated upon*. It is based on a simple principle which I claim to be original, as not having been previously enunciated, viz.:—

When a stock has developed queen cells in preparation for swarming, a broodless artificial swarm can be made from it if some or all of its queen cells are sealed and this swarm will behave in every way as a natural swarm; but if none of the queen cells are sealed such an artificial swarm will continue to prepare for swarming.

If, then, we ascertain by weekly inspections when a stock develops queen cells and calculate when these are likely to be sealed, we can operate on it as described below and so obtain a reinforced natural swarm which should give a good account of itself.

In the early part of the season the stocks are worked as indicated in steps 1–5 on p. 57. From the time when pear and plum blossoms open, the stocks should be visited regularly once a week until mid-July and inspected for queen cells. This is done by raising the upper (shallow) brood box to the position shown in the illustration. (Plate VII, Fig. 1.) If queen cells are present, most of them, if not all, will be seen at the bottom of the combs in the upper box and sometimes others between the top bars of the lower box. Calculate how many days will elapse before the oldest cells will be sealed, remembering that there are only five days between the hatching of the egg and the sealing of a queen cell.

When closing the hive, use a little smoke, especially along the line where the boxes meet, to ensure that the queen is out of danger. If no queen cells are seen it can be assumed that there will be no swarm before the next visit.

Note that there is no withdrawal of combs and very little disturbance of the bees, and that a glance is sufficient to show whether the bees are going to swarm or not. When two persons are present, one to lift and the other to inspect, the inspection can be completed within three minutes—that is to say, of about twenty stocks in an hour.

It may be noted that this method of inspection is always reliable when the upper brood box is shallow. In the case of a double brood chamber it is usually, but not invariably, reliable.

If queen cells are found it is well to decide whether they are

(1) In very early stages.
(2) Near to being sealed, or sealed.

If (1), defer manipulation until cells are near to sealing. If (2), place under the whole stock a framed sheet of excluder, of a reliable pattern, and raised in front by the 1″ blocks. (Plate VII, Fig. 2.) Bees, but not the queen, will be able to emerge from the hive over the whole area of the excluder, and at the sides as well as the front of the hive. Make sure that they cannot leave any part of the hive without passing through the excluder. Within three or four days the queen cells will have been sealed and the bees will have swarmed and returned, perhaps several times. If the weather has been bad they may not have attempted to swarm but will still have a strong impulse to do so.

If an artificial swarm be now made from them they will behave as though it were a natural swarm.

Proceed as follows:—

Lift the whole stock from its floor and screen to a temporary stand a few feet to the rear of its original position. (Plate VII, Fig. 3.) On the screen and floor, which remain in their original positions, place a fresh brood box prepared to receive a swarm; that is, containing two combs with stores, if available, some empty combs and frames of foundation *but no eggs or brood*.

The next step is to find the queen in the parent stock. After a few puffs of smoke at the base of the hive, and after a short interval, she will most likely be in the upper (shallow) box. Place the comb in which she is found in the prepared hive on the orignal stand and cover it at once. After a short time many flying bees will have returned home and queen and bees will be quiet and contented. Now gently shake bees and queen from the comb on which she was introduced and restore it to its place in the original hive. No brood or eggs should be left to the swarm.

The supers, with bees, may be transferred to the swarm at the next visit. The screen is left under the swarm for a few days. (Plate VII, Fig. 4.) It should be removed when the queen is found· to be laying normally. Supers should be added as needed.

At the next weekly visit the original stock containing the queen cells may be placed over a Snelgrove Board. (Plate VII, Fig. 5) above the supers, and the swarm reinforced (pp. 25, 26) after a young queen has become fertile. Alternatively, the original stock may be moved to another position as an additional stock within a week after the young queens have emerged; or, of course, it may be divided into nuclei for queen-rearing. The swarm will receive its flying bees.

If the beekeeper desires to avoid a search for the queen he may have recourse to the "Shook Swarm" Method described on p. 69. In this case there will be no need to alarm the bees as recommended by Doolittle. (Combs with queen cells, if these are needed, should not be shaken, but the bees should be disloged from them by means of a feather.)

If young queens are needed, one or more nuclei, each with a queen cell, may be formed before shaking. The remaining brood, minus queen cells, may be used to reinforce other stocks.

Manipulations such as these are best performed *during the evening*. The excluder-screen may be left under the newly-made swarm for about a week but not longer. This will not result in any mortality of the drones. About half a dozen excluder screens are sufficient for managing fifty hives.

By this procedure a strong natural swarm, reinforced by all the flying bees, will be formed which will give a maximum yield of honey for the season.

Important points to remember are:—

(1) The excluder screen should be inserted before and near to the day when queen cells are sealed, or immediately, if any are found to be sealed.

(2) On no account should an impending swarm be detained by a small strip of excluder at a hive entrance. This would be choked with drones.

I have described this method in considerable detail for the benefit of beekeepers of little experience. Experienced beekeepers will not need more than the following simple directions:—

(*a*) When queen cells near to being sealed are found, insert the detaining screen.

(*b*) Three or four days later, when the bees will have attempted to swarm, or have been prevented from doing so by bad weather, make an artificial swarm, including the queen, on the original site, and over the detaining screen. At the next weekly visit examine the swarm to ascertain if the queen is laying, and transfer the supers, with bees, to it.

(*c*) Use the original stock for reinforcement by the Heddon method (p. 97), or by placing it over a Snelgrove Board (Pl. VII, Fig. 5), for making nuclei.

ADDITIONAL OBSERVATIONS.

If two or more supers are in use it is advisable to remove them before inspecting for queen cells.

If the brood nest has been augmented by a third (shallow) brood box (p. 57), both spaces between the boxes should be inspected for queen cells.

A detaining screen is easily made by pinning an excluder to a frame such as is used in the construction of a Snelgrove Board (p. 25) *without wedge openings*. If stocks are judiciously managed about half a dozen detaining screens will be sufficient for an apiary of forty to fifty stocks, as only a small proportion will prepare to swarm and these not all at the same time.

The reason for waiting till queen cells are in an advanced stage is that if a detaining screen is inserted when they are in an early stage the bees will sometimes break them down, only to start a fresh lot a little later. This occasionally happens also if very cold weather sets in, whether a screen is used or not.

CHAPTER IV

OTHER METHODS OF SWARM PREVENTION

1. Destruction of Queen Cells.

It is well known to all beekeepers that swarming can be delayed by the destruction of queen cells, and most novices on beekeeping soon arrive at a stage in their experience when they attempt to prevent it altogether by this means. They are sometimes successful but very frequently they meet with disheartening failure. It is a popular notion that if the queen cells are broken down once every seven days there will be no swarms. This, however, involves a serious fallacy. Normally the young queens are reared in newly-made cells in which the mother queen deposits eggs. Nine days later the fully developed larvae are sealed in their cells and within a few hours after this the swarm issues. If the first batch of cells be destroyed the bees straightaway proceed to rear queens in others and without waiting for the queen to deposit eggs in specially built cells they will select ordinary worker cells containing eggs of one, two, or three days old or even those containing young larvae. Indeed when the swarming impulse is strong in them and they have been made desperate by frequent interference on the part of the beekeeper they can produce queens from larvae of comparatively advanced age. If for example they select a larva which has just hatched from an egg (fourth day) the resulting queen cell will be sealed five days later and the " every seventh day " beekeeper will lose his swarm. When we consider that a queen larva is specially fed as such from the fifth to the ninth day after the egg is laid it is obvious that it is possible for the bees to produce sealed queen cells four days after the beekeeper has destroyed all the cells he can find. On the assumption therefore that queens would be reared from larvae of one to two days old the destruction of queen cells every five days would appear to be necessary to prevent swarming.

Any method which depends on the breaking down of queen cells is undesirable. Not only is much labour involved on each occasion,—labour to be repeated several times, but the difficulty of detecting the queen cells without missing a single one is easily realised. Some may be in the earliest stage and hardly discernible, others small and embedded amongst worker brood,—their protruding ends easily mistaken by a beginner for isolated drone cells. Some may be covered by the crowding bees and therefore be passed unnoticed. Indeed to make sure that none is missed it is necessary to shake most of the bees from each comb before examining it. Apart from the time and labour involved for the beekeeper how great is the disturbance to the bees and how many hours must elapse before they have fully resumed their various duties in the hive.

Some people imagine that queen cells must be cut out with a knife. In this way they destroy neighbouring brood and mutilate the combs, leaving large holes, which when repaired are filled with drone cells. It is sufficient to crush each cell with the finger or a lead pencil, but however it is done it is a rather distasteful business.

This crude method of swarm prevention has other defects. Most bees, and especially those prone to swarming such as the Dutch and Carniolans, become impatient of the repeated destruction of their queen cells, and may swarm before these are sealed, or even without them. Other bees, when forcibly prevented from satisfying the swarming impulse, become listless, and cease to work actively in the fields.

2. THE DEMAREE METHOD.

This is one of the most widely practised methods of swarm prevention. It was devised by a Mr. Demaree of the U.S.A. in 1892, and is still popular in that country and to some extent is used in the British Isles. It is often imperfectly described in bee literature with the result that many failures occur on account of the neglect of essential details. In the hands of a skilful beekeeper it not only usually prevents swarming but often results in excellent yields of honey. Its principal defects are that it involves much labour in the frequent manipulation of large stocks and the breaking down of queen cells, is unsuitable for sections, and there is some risk of swarming throughout the season.

The underlying principle of the method is as follows:—If brood is removed from the vicinity of the queen and placed over an excluder in the part of the hive remote from her, whilst at the same time she is given ample additional laying room, swarming is discouraged.

Procedure varies in the hands of different beekeepers but generally in this country it is as follows:—

The stock is developed in the spring in the usual way until it occupies at least two brood boxes. The combs are then separated, those containing brood being placed in one box and those without brood in the other. The queen and one frame of young brood are placed in the latter. An excluder is then placed over the box containing the queen, a super above the excluder, and the box of brood above the super. The queen continues to lay without interruption, the bees pass freely through the super, and those in the box of brood at the top begin to construct queen cells as they would do if they were queenless. After seven days the whole hive is examined again, queen cells are broken down, and combs from which the bees have "hatched" in the top box are exchanged for others containing fresh brood from the bottom box. Queen cells may again be built in the top box and therefore the operation is repeated until the risk of swarming is past. Some beekeepers do not continue to exchange the combs, in which case the top box is gradually filled with honey whilst the nurse bees descend to the box containing the queen,—a condition considered to be conducive to the development of the swarming impulse.

I have several times read descriptions of this method which contained no reference to the necessity of destroying queen cells and have known beekeepers who have come to grief in consequence. If the cells are not destroyed by the beekeeper, or, in certain circumstances, by the bees, a young queen will be bred in the top box, and finding herself imprisoned may force her way through the excluder and depose the reigning queen below.

It is useless to attempt this method on a stock which has already developed queen cells in preparation for swarming. In such a case the bees will continue to rear young queens from eggs in the lower box and will ultimately defeat the "every seventh day" beekeeper.

When the main honey flow sets in it is unnecessary to continue to raise the brood combs from the bottom box. The brood hatches out in the top box and gives place to honey, supers being added as necessary. Some beekeepers use only standard brood combs in connection with this method but this involves troublesome work in lifting.

When sections are desired it is recommended that the brood combs be raised until they fill two brood boxes above the excluder. The bees are then shaken from the lower one and a rack of sections is substituted for it. This crowds the bees in the sections and so increases the probability that these will be drawn out and filled.

Apart from the heavy work incidental to this method there are other disadvantages which cannot be disregarded. Honey extracted from combs which have contained brood in the same season contains more or less pollen, the flavour of which, easily discernible to a good judge, makes the honey inferior to that obtained from supers entirely separated from the brood. Sections placed immediately under the top brood box are likely to be spoilt by occasional cells filled or half filled with pollen. These cells are sometimes completed with honey and sealed over like others, and are not detected until they produce an unpleasant flavour in the mouth of the consumer. Moreover sections placed between two brood boxes will be travel-stained if they are left too long in the hive.

The Demaree plan does not ordinarily provide for requeening. It is possible however to allow a young queen to "hatch" in the top box and to fly for mating from a hole bored in the side of the hive. Needless to say she must not be left there very long after becoming fertile, otherwise she will displace the stored honey by brood.

3. THE "HAWKINS" METHOD.

This method has been successfully practised for many years by Mr. E. G. Hawkins, of Frome. In the hands of a skilled beekeeper who has adequate time, and does not object to frequent manipulations and heavy work, it prevents swarming and often results in large yields of honey. It is carried out as follows:—

When the bees fully occupy the first brood box a second is added as a super, no excluder being used. Later a third box is added and others as necessary. On the approach of the swarming season the hive is thoroughly examined every seven days and the combs re-arranged. Those containing young brood are moved to the lowest box, those containing honey to the top box, and sealed brood is placed in an intermediate position in the middle box. Every week the process is repeated, the honey being raised and the young brood lowered. At the same time queen cells are diligently sought for and destroyed when found. The queen is able to roam throughout the hive but naturally remains in the neighbourhood of the youngest brood, while the honey is stored in the top boxes, taking the place of brood as it emerges. The stock grows apace and in a good season several boxes of combs will be occupied and much honey stored.

As will be realised it is no easy task to search frequently for queen cells in a hive consisting of several full-sized boxes, and the overlooking of a single queen cell frustrates the whole plan.

Honey must be removed from the upper boxes as soon as convenient, for when the flow is over, or if bad weather intervenes, the brood nest gradually ascends and the surplus honey is converted into brood.

Notwithstanding the successful results often obtained by this method its chief defects are that queen cells must be searched for, destroyed, and honey extracted from brood combs, whilst there is no provision for requeening, and the amount of time and labour involved is very great. It is, of course, unsuitable for the production of comb honey.

4. REMOVAL OF QUEEN.

We occasionally see statements in English bee literature to the effect that swarming may be prevented by the removal of the queen and the destruction of all queen cells but one. No distinction is made between stocks preparing to swarm and those which are not, and detailed directions as to when the cells are to be broken down are usually vague or omitted. In my own experience this procedure has many times proved unreliable. In the case of a stock which has already developed queen cells the swarming impulse persists until the young queen is ready for her mating trip and often

the swarm leaves the hive with her. The beekeeper
may be sure that he has left only one queen cell but
he must not forget that queenless bees with the swarming
impulse will often persist in raising additional queen
cells whilst they have larvae on which to build them.
If this happens the swarm leaves when the first queen
emerges. When the bees have not previously prepared
for swarming the plan is likely to be successful provided that

(1) Only one queen cell is left. This should be
     selected before it is sealed as cells already
     sealed sometimes prove to be empty. The
     selected cell should not be amongst the drone
     brood.

(2) Other queen cells are broken down on the fifth
     and tenth days after the removal of the queen.

(3) The bees do not accompany the young queen on
     her mating flight.

5. REMOVING THE QUEEN AND REQUEENING.

Hutchinson, in his book on *Advanced Bee Culture*
(page 73), says:—"Probably the only *certain* method
that has been used to any extent in this country (U.S.A.)
is that of removing the queens just at the opening of
the swarming season, leaving the colonies queenless
about three weeks. Of course the queen cells must
be cut out at least once during this interval. Although
a few good men practise this method, I could never
bring myself to adopt it,—there is too much labour."

This method has the advantage of limiting the
amount of brood to be fed during a succeeding honey
flow, but to ensure success the queen cells must be
destroyed on the fifth and tenth days, thus increasing
the labour of which Hutchinson speaks. There is a
risk, however, in keeping bees queenless for three
weeks during the summer season, for some strains of

bees quickly develop laying workers when they have no young larvae and are deprived of the means of rearing a queen. In 1927 I tried this method on about 30 stocks with little success. Six stocks headed by imported American Italian queens all had fertile workers in the third week. My stocks of black bees escaped this pest.

6. REMOVING THE QUEEN AND REQUEENING.

A modification of the last method is described by G. S. Demuth and is favoured by some people. The queen is removed and after the queen cells have all been destroyed a new queen is introduced on the tenth day. This involves labour and also the provision of a new fertile queen at a given time. Few English bee-keepers have young queens fertilised sufficiently early in the season and consequently purchases must be made and the trouble of introduction undertaken. If the bees have previously prepared to send forth a swarm the newly-introduced queen will in all probability issue with it. In the case of stocks which have not made swarming preparations the method is successful, but it should be considered as delaying rather than preventing swarming. If the new queen is a prolific one and variable weather interrupts the ensuing honey flow the conditions conducive to swarming will be produced in the stock later in the season.

7. DEQUEENING.

As an improvement on the two preceding methods I will now describe the following one devised by myself in 1927 and given to the Somerset Beekeepers' Association at its annual meeting in Bristol in 1929. Although it has the disadvantage of involving the destruction of queen cells it works well whether the bees have developed the swarming impulse or not. It does not involve extra equipment but allows for requeening from a good stock and increases the yield of honey.

I had been endeavouring to keep bees queenless during the greater part of the honey flow on the ground that young brood during that period not only resulted in bees which would take no part in surplus storing but absorbed a large amount of food which might be economised in the form of honey. Method 5 proved troublesome in respect of laying workers and in Method 6 the queenless period was too short. A simple modification of both proved satisfactory.

A stock is given ample room in advance without the use of queen excluder, and if possible is carried on in this way without swarming until a week or ten days before the period of usual honey flow. The queen is then removed and queen cells are broken down on the fifth and tenth days as in methods 5 and 6. To encourage the bees to continue their field work without interruption the removed queen may be caged over the feed hole until the tenth day when she is removed altogether. After the queen cells are destroyed on this day the bees have no means of raising further queens. They are now given a very small piece of comb containing about a dozen freshly laid eggs taken from the beekeeper's best stock. This small piece of comb, cut preferably from a newly-built comb of the same season, is grafted into one of the combs of the stock to be requeened. This comb is put into the centre of the brood box and the latter covered with an excluder, over which are placed the supers. The bees will soon begin to construct queen cells over the eggs and finally one young queen will be mated and head the colony. She will not be laying however until at least five weeks after the removal of the original queen. During the whole of this time the bees are never without queen cells or a queen and so they keep in good heart. Laying workers do not make an appearance in these circumstances. By the time the first young queen is hatched there are no young nurse bees, the honey flow is over, and swarming is out of the question. During the honey flow most of the bees become field bees and few are needed for nursing, the result being that most of the honey gathered is stored in the combs.

Care must be taken to ascertain whether the young queen "hatches" and is safely laying. Should she be lost and the stock left queenless for a time fertile workers may be expected.

I have on several occasions by means of a brush and a little melted wax fastened the little piece of comb containing the eggs into a small glass jar and inverted this over the feed hole. There the bees can be watched as they construct the queen cells. These should be placed below the excluder when sealed. There is no need to reduce them to one unless the beekeeper considers that the one is better than the rest.

Very good crops of honey have been obtained by this method. Its disadvantages are the labour involved, and the risk that the young queen may fail to become mated.

8. CLIPPING THE QUEEN'S WINGS.

The practice of cutting the queen's wings on one side of her body to prevent her from decamping with a swarm in the owner's absence is widespread and has been often advocated in bee literature. Some beekeepers clip the wings of all their queens in the spring and then feel secure against the loss of swarms. But the plan seldom works so well and usually involves many difficulties and disappointments.

The queen's wings are clipped before the arrival of swarming time. This can be done by holding her gently by the thorax between thumb and finger and carefully snipping off with a fine pair of scissors at least half of the two wings on one side of her body. In this condition she is unable to fly, and when she issues with her swarm at the normal time, instead of rising into the air with the bees she drops to the ground in front of the hive, crawls excitedly about making vain attempts to fly, and finally comes to rest a few yards away where her presence may be indicated by a little group of the bees which have detected her. Meanwhile the swarm, having lost her, returns to the hive.

If the beekeeper is present all may be well. He quickly removes the hive to a new position and puts a prepared hive in its place. The returning swarm enters this and the queen, when found, is run in with it. But if it is to be assumed that the owner must be present daily he may as well be content with natural swarms and avoid the trouble associated with wing clipping.

Should he be absent, however, there will be no external indication a day or two later that a stock has swarmed. The queen will probably have perished, and only a thorough examination of his stocks will enable him to ascertain what has happened. Queen cells will warn him that within a few days his swarm will come out again, but this time accompanied by a young queen that can fly. As a rule he destroys all queen cells but one, taking the risks that the one may be "empty", that he may miss an inconspicuous one, or that the bees will make others as soon as his back is turned. And so he often loses his swarm,—sometimes without knowing it. The "clipped wing" method is unsuitable for an out-apiary unless this is frequently visited, and in any case it demands constant vigilance, frequent manipulations, and considerable resourcefulness on the part of the beekeeper. Moreover the clipped queens if found and hived are rather liable to be superseded before the following season.

9. "SHOOK" SWARMING.

This picturesque title is applied to a method said to be widely used in the U.S.A. When successful it is quite suitable to our own conditions of beekeeping. It is described by both Hutchinson and Doolittle, who however do not fully agree as to its application. The former considers it should be used on stocks that have made queen cells in preparation for swarming, whilst the latter prefers that it be applied prior to the advent of the swarming impulse (Chap. III).

Essentially the procedure is as follows:—

(a) When no preparations for swarming have been made:—

Shake all or most of the bees from a strong colony in front of a newly-prepared hive filled with broodless combs, or, as Doolittle insists, combs filled with honey recently gathered from fruit bloom, etc. One comb containing a little unsealed brood is placed in the centre of these to encourage the queen to continue laying without interruption. The supers are then placed in position over an excluder and the hive completed.

If all the bees are shaken the brood may be utilised elsewhere; if not it may be made to provide future reinforcements of field bees by the Heddon method (vide page 97).

To avoid any chilling of the brood this may be placed for a short time on the old stand to catch some flying bees. It is then moved to a new position whence the flying bees will return to the swarmed stock, now on the old stand. As they return young bees will be "hatching" to take their places.

There is no reason why this method should not always be effective, but if used too early the swarming impulse develops later in the season.

(b) When the bees have prepared to swarm:—

In this case the conditions of natural swarming are simulated as far as possible. Before the shaking is begun the hive is jarred and pounded upon for a short time so as thoroughly to alarm the bees. They at once fill themselves with honey as they do prior to swarming. When they have entered the new hive they find themselves in the exact circumstances of a newly-hived swarm. The hive should be completed and the brood dealt with as described in the preceding paragraphs.

No young brood must be placed in the centre of the prepared hive for, as is reasonable to suppose, and as Doolittle actually found in some cases, this would encourage the bees to continue their queen cell building and subsequently to swarm.

I must speak with caution about this method, having found on a few occasions that the bees have swarmed and decamped the day after being shaken (vide p. 61).

10. HAND'S METHOD.

Mr. J. E. Hand, an American beekeeper, prevents swarming by the use of two hives on a double stand, the latter being ingeniously fitted with a series of entrances and "switch levers" which enable him to divert the field bees into one or other of the hives at will (vide "*ABC and XYZ of Bee Culture*," 1913 Ed., p. 48).

When a stock has become strong and is occupying supers, the queen and a frame of brood are taken out and transferred to the neighbouring hive on the same stand. This hive is filled up with empty combs or foundation, and by the movement of one of the levers the field bees, still using their original entrance, are switched over to join the queen. The supers are then transferred to this new stock.

The original stock, now queenless, is provided with

a virgin or young fertile queen, and uses a new entrance at the back of the hive. Eight or ten days later the flying bees are again diverted to the new colony which now becomes the main stock, and a side entrance is made available for the original stock.

In this way swarming is usually prevented. Mr. Hand relies on the youth of the queen given to the original stock to prevent it from subsequent swarming. If the newly-made stock shows signs of swarming the field bees are switched over to the original stock to which the supers are then returned.

This method is sound in principle but is obviously rather cumbersome and involves the use of additional equipment and the provision of an extra queen. If the original stock were allowed to rear its own queen a third abstraction of field bees would be necessary just before the young queens "hatched", but for this the double floor does not provide.

11. SIMMINS' METHOD.

Mr. S. Simmins, one of the most able and original of English beekeepers, discourages swarming by arranging that the bees are engaged throughout the swarming season in comb-building either underneath the brood nest or in a space between the brood nest and the hive entrance. They are never allowed to complete the combs in these positions for they are removed when partially built and transferred to the supers. In a hive of ordinary dimensions a rack of shallow frames provided with starters of foundation is placed below the brood nest. In long hives the combs comprising the brood nest and the supers above it are kept towards the back of the hive, and between them and the entrance are placed five or six brood frames provided only with starters. As fast as the starters are drawn out into combs they are cut up, fitted into sections, and transferred to the supers. If any contain eggs these are destroyed by exposure to ordinary temperatures for three days.

The provision of newly drawn combs in the sections is a great inducement to the bees to store in them, and the constant occupation of the wax-secreting bees is a strong deterrent of swarming.

The chief drawback to this plan is that the bees refuse to work on the starters if little or no honey is coming in, and there is then no inhibitory effect on swarming. In any case the method implies frequent manipulations and much labour.

12. DETENTION OF QUEEN.

Methods of swarm prevention which depend upon the forcible detention of the queen, and incidentally of the drones, *unless only temporarily*, by means of queen excluder are to be deprecated.

Amongst the crudest and worst of these is one which has been widely advocated in recent years in connection with the sale of certain excluders. The beekeeper is advised to place a full-sized sheet of excluder under the brood nest so that only worker bees can leave the hive. It is not difficult to picture what happens. The drones make desperate efforts during the whole of the day to escape from the hive, and in so doing often worry themselves to death. Some get partially through the spaces and die there, being unable to advance or retreat. If they are liberated at intervals by the bee-keeper they must either perish outside the hive or be re-admitted only to experience the same torture again. If they are numerous they greatly impede ventilation and the passage of the worker bees. When the time for swarming arrives the swarm issues as usual and the queen makes frantic efforts to follow it. Sometimes she succeeds in forcing her way through the excluder and if the beekeeper loses the swarm he deserves it. If the queen cannot escape the swarm returns to the hive, and the effort to swarm is renewed on the next and succeeding days. In a week's time after the first attempt young queens will emerge from their cells and one of these is likely to force her way through the excluder and get away with the swarm. If not, she kills the old queen, and being unable to take a mating flight becomes a drone-breeder and ruins the stock. When the beekeeper discovers that a swarm has attempted to leave he may of course deal with the hive in several ways so that further attempts are delayed, but it is most difficult to destroy the swarming impulse in such a stock. If baulked in frequent attempts to swarm the bees ultimately become disheartened and refuse to work in a normal way. On the other hand, neither queen nor drones will come to grief if the whole surface of the excluder is exposed for ventilation and passages and retention does not exceed about ten days.

13. Use of Swarm-Catchers.

Many ingenious devices have been invented for the purpose of trapping swarms as they issue. So rarely however are they seen in use that this fact alone testifies to their ineffectiveness. The principle of them all is the same.

A swarm-catcher is essentially a box of which the sides are sheets of excluder zinc. When this is fitted accurately to a hive entrance the worker bees can leave and return to the hive only by passing through the catcher. The queen and drones cannot pass through the excluder at the hive entrance to enter the catcher, but when eager to do so they search until they find one or more specially devised escapes, which, while permitting them to enter the catcher prevent them from returning to the hive. Once in the catcher they are prisoners. The catcher is usually provided with some empty combs for the reception of the swarm.

When the swarm issues the bees stream out through the catcher and circle in the air in the usual manner. The queen tries to accompany them, enters the catcher and is trapped in it, where she is ultimately joined by the swarm. The catcher is then removed by the beekeeper and the swarm hived in the usual way.

Unfortunately excluder zinc cannot always be relied upon to intercept a queen. If a single bee-space in the excluder be slightly widened through buckling or otherwise, or if the queen be small, she will force her way through. The slightest carelessness in fitting the catcher may allow her to escape. Sometimes she does not even enter it and returns to the hive.

Quite apart from the possibilities of failure in its use, a swarm-catcher, when applied to a hive for any length of time in anticipation of a swarm that may issue, is an instrument of cruelty to the drones which daily become trapped in it and after fruitless efforts to escape come to an untimely end.

It may however prove useful in an apiary where swarming is allowed, for if it be quickly and gently applied to a hive during the actual time of swarming it may be the means of catching the queen, and therefore the swarm, for the queen often leaves the hive after, and not before the issue of the majority of the bees.

14. THE "PECK" METHOD.

During the last two or three years a method practised by Mr. S. Peck, of Histon, Cambridge, has been widely recommended in this country. Its success depends upon two principles, viz.: the periodical re-arrangement of the brood nest and the detention of the queen by means of queen excluder.

The bees are worked in a single brood box throughout the summer, and this is supered in the ordinary way. The brood box contains twelve standard frames and is rather longer than usual to allow for some lateral movement of the frames during manipulations. Early in the season, usually during April, the brood chamber is divided into two compartments by means of a queen excluder partition. One of the compartments, sufficient to hold five combs, is covered with a half sheet of queen excluder and the portion of the entrance leading to this compartment is also stopped by a strip of the queen excluder. This compartment therefore is so enclosed by the queen excluder that when the queen is in it she cannot possibly come out with a swarm.

Early in the season five combs containing the least brood, together with the queen, are placed in this compartment, the remaining combs being in the other compartment. Supers are added as usual and placed over both compartments.

Ten days later the hive is overhauled. The five combs in the compartment occupied by the queen are now practically filled with brood. The comb with the least amount of brood on it is left in the middle of the compartment, the bees are shaken from the other four combs, and these are exchanged for four others, with their adhering bees, from the neighbouring compartment. If queen cells are found they are broken down.

Ten days later the process is repeated, four combs of brood from the queen's compartment being exchanged for four broodless combs from the other compartment, and queen cells, if any, are again broken down.

The process is continued every ten days as long as necessary. It will be seen therefore that every tenth day the queen receives almost broodless combs in which to continue laying, and at the same time most of the nurse bees migrate through the excluder to look after the brood in the outer compartment.

This periodical re-arrangement of the brood nest in itself tends to some extent to inhibit the desire to swarm. If a swarm issues between two visits of the beekeeper it returns to the hive within a few minutes, and it is said the bees then break down the advanced queen cells. Other queen cells are destroyed by the beekeeper at the next examination of the hive.

Although this method may involve several important operations on a hive in a single season, and therefore considerable labour, it is said to work satisfactorily; supers are readily occupied and filled, and one person can deal with many hives in a day. An important defect is that the queen and drones are confined to the hive by means of queen excluder. This defect however is not so serious as in the case of method 12 because attempts to swarm are less frequent and imprisoned drones are liberated at the periodical examinations.

Mr. Peck arranges for requeening in the case of a failing queen either by the introduction of a young fertile queen or by allowing a queen cell to "hatch" in the queen compartment. In the latter case he does not allow the young queen to leave the hive for a mating flight until the old queen has been superseded. This of course postulates care and judgment on the part of the beekeeper.

(15) In the *Bee World* of October, 1931, P. Risga, of the University of Latvia, described a method somewhat similar to that of Mr. Peck. A larger brood box is used and this is divided into two compartments by means of a partition made of queen excluder. The brood from the queen compartment is periodically exchanged for broodless combs from the other compartment, but there is no imprisonment of either queen or drones by means of an excluder at the hive entrance to the queen compartment.

(16) Mr. G. S. Demuth, editor of the American Journal *Gleanings*, an eminent and reliable authority, states that swarming may be prevented by taking from

a stock unsealed brood and substituting for this combs of sealed brood from other stocks. (*Gleanings in Bee Culture*, August 1929.) He gives the following as a method of swarm prevention: —

Shake all the bees on to two broodless combs after having first removed the queen, and leave them queenless and broodless for a whole day. Fill the hive with combs of sealed brood and introduce a young queen.

The young brood can be cared for by another colony and when sealed can be used for further substitutions. It would seem likely that swarming conditions might recur if the substitution of sealed for unsealed brood were not repeated.

(17) In contrast to Mr. Demuth, Mr. Jay-Smith, one of the most experienced of present day beekeepers in the U.S.A., states that swarming may be prevented by adding brood to a strong stock. (*American Bee Journal*, June, 1925.) This is feasible and in accordance with the Gerstung theory, for the addition of unsealed brood would at once give occupation to any excess of nurse bees. It seems likely however that as in the case of the preceding method the inhibitory effect would be temporary only unless the additions of unsealed brood were continued.

(18) Mr. W. Herrod-Hempsall, adviser in Beekeeping to the Ministry of Agriculture, consistently recommends as the best means of swarm prevention the breeding of queens only from stocks which show a non-swarming tendency. That this tendency varies considerably in stocks of the same race is well known. In an apiary of any size there is usually a small proportion of stocks which do not swarm under natural conditions in a particular season. This variation from the normal behaviour is considered to be inherent in the queen and her progeny and to be hereditary. By constant selection

of both queens and drones from such stocks it is possible
to develop a strain in which swarming is reduced to a
minimum.

Selection of queens from non-swarming stocks is
easy but the selection of drones is often a difficult matter.
Although it is believed that queens are likely to be
mated by drones from their own hives, the well known
difficulty of maintaining a pure race and the prevalence
of hybrids prove that they often mate with drones from
neighbouring apiaries. To minimise the risk of cross-
mating Mr. Herrod-Hempsall recommends that drones
from a selected stock be introduced to the nuclei containing the young queens before they fly to be mated.
This can best be done by introducing "hatching"
drone brood to the stock from which the nuclei are made
a fortnight or more before the drones are needed for
mating. Alternatively the mature drones may be
introduced directly and will be readily accepted in the
nuclei.

CHAPTER V

ARTIFICIAL SWARMING

THE subject of swarm control is intimately bound up with that of artificial swarming. Indeed the principal method described in this book really amounts to the making of a strong artificial swarm within the hive.

Whilst all beekeepers are concerned with the problem of controlling natural swarming many are desirous of increasing the number of their stocks, and in the absence of natural swarming, of effecting this by artificial means.

Methods of dividing stocks for increase are innumerable, and few aspects of beekeeping afford greater scope for ingenuity or demand more prudence on the part of the beekeeper. There are certain cardinal principles which must be observed if success is to be obtained, and it is important that the beekeeper be familiar with them.

In the first place it must be recognised that the division of stocks before the honey-flow, whether by natural or artificial swarming, usually results in a diminished yield of honey, the amount stored as surplus depending chiefly on the ratio of foraging bees to those engaged in domestic duties. The more powerful the stock the greater is the proportion of bees available for field work. Such a stock divided into two colonies would have two brood nests, and therefore almost twice the number of bees would be needed to perform the necessary domestic duties.

Division should be effected only while the weather is warm, and if possible while the bees are still gathering honey. If queens are to be reared for the new colonies the incréase must be effected sometime before the drones are likely to be expelled from the hives, and ample time must be allowed in which an artificial swarm may develop a good brood nest before cold weather sets in. A great deal of time can be saved and success made more certain if young fertilised queens are available for immediate introduction to new colonies. Feeding is usually necessary to stimulate the queens to lay and to ensure adequate stores for winter. Only strong stocks should be swarmed artificially, and if young queens are to be raised provision should be made that this be done by the young and not by the old bees.

If a beekeeper who desires artificial increase is prepared to sacrifice some of his prospective honey crop the best time to divide a stock is undoubtedly the natural swarming season. At that time it is possible, by skilful manipulations, to build up from one stock two or more good colonies which will yield surplus honey, or several which will develop into stocks sufficiently strong to go through the winter.

If the crop of honey is the primary consideration, however, artificial swarming may be left until the close of the main honey flow, but at that time it must be restricted to moderate limits, and generous feeding will be necessary.

It must be remembered that when an artificial swarm is made by division of combs so that brood combs have to be occupied by returning field bees the operation should be carried out when the bees are flying strongly. Artificial swarming by division is not advisable in the case of stocks which are building queen cells in preparation for natural swarming.

If it is desired to give a ripe queen cell to a newly-formed nucleus or a colony recently dequeened this should not be done until the queenless bees have begun to build their own queen cells, i.e. two or three days after they have experienced the loss of their queen. Cells inserted earlier are often broken down.

When a stock or nucleus is placed on the stand of another stock to receive its flying bees the latter do not, as a rule, attack the queen surrounded by her own workers. When honey is coming in no precautions against an attack are necessary. In times of scarcity, however, and especially when robbing is prevalent, it is prudent to cage the queen for two days. This is conveniently done by means of a pipe-cover cage pressed into the comb on which the queen is found.

If a stock or nucleus is to be transferred to a new position in an apiary while the bees are active the loss of its flying bees can be entirely avoided by its removal for a few days to a distance of two miles or more from the apiary. On its return to the apiary the bees will accept any position as their new home.

François Huber (1750–1831) found that by placing a partition in the middle of a stock of bees and providing an entrance for each part he obtained two stocks, one headed by the original queen and the other by a young queen. Such a method of increase, while satisfactory for Huber's observations, would be considered crude to-day, and would not lead to a good result. The bees on the side of the partition remote from the queen might raise queen cells but it would be at least three weeks before a young queen would be laying and a further three weeks before any of her young bees would "hatch". By this time the little stock would have so diminished in numbers as to be of little or no use.

The following are descriptions of various methods of effecting artificial increase:—

1. To make Two Stocks from One.

From a strong stock take one frame of brood with the adhering bees and the queen, and place it in the centre of a prepared hive. Fill the hive with empty combs or sheets of foundation. Remove the parent hive to a new position several yards distant and put the prepared hive in its place. The field bees will gradually return to their old location and a good stock, headed by the queen, will be formed. If this is done sufficiently early this stock will be likely to store honey in supers.

The parent colony containing the young bees will rear a young queen but it will be a great advantage if a young fertile queen be at once introduced to it.

2. To make Three Stocks from Two.

From No. 1 stock take four or five frames containing brood of all ages, eggs, and stores. Shake the bees from these and put them into a prepared hive. Fill up both hives with empty combs or sheets of foundation. Remove No. 2 stock to a new position some yards distant and in its place put the prepared hive. This will receive the field bees from No. 2 stock and these will rear a young queen and eventually become a strong colony. In this way one stock provides the brood and the other the bees which go to make up the third stock. Much valuable time is saved, however, and the field bees will not have the task of queen-rearing, if the new stock be provided at once with a laying queen. She should be caged for two days before liberation.

3. To make an Artificial Swarm from Several Stocks.

Assume that five good stocks are available.

Select four stocks and from each take two frames containing brood and stores. Shake the bees from them and place them in a newly-prepared hive. To each hive, including the prepared one, add two empty combs or frames of foundation. Move a fifth stock to a new location and place the prepared hive on its stand. The field bees will return to this hive, occupy the brood, and subsequently raise a young queen. If a queen is available for immediate introduction to this hive so much the better. She should be caged for two days.

4. To Double the Number of Stocks.

Suppose there are five stocks and it is desired to make them into ten.

During the first or second week in May or as soon as the strongest stock (No. 1) fills its hive, and before it makes preparations to swarm, remove a nucleus colony from it in the following way:—

Take out three combs of bees, including the queen, and place them in a hive. One of these frames should contain ripe sealed brood, another unsealed brood, and the third should be well provided with stores. Into this nucleus colony shake all the bees from one additional well covered comb. These extra bees are to compensate for those which will leave the nucleus to return to the old hive on the ensuing days. The nucleus colony should be warmly covered, placed on a new site, and fed with syrup.

Cover the parent stock warmly and feed with syrup. Do not insert foundation in place of the removed combs at this stage, unless honey is coming in fast. The stock will now begin to raise queen cells.

Nine days later make a nucleus colony from each of the stocks Nos. 2, 3, 4, and 5, in the manner described above.

Three days later supply each of the parent stocks Nos. 2, 3, 4, and 5 with a ripe queen cell from stock No. 1, taking care that at least one good cell is left in No. 1. These queen cells will be near " hatching ". Cut each off with a piece of the adjacent comb and let it hang between the combs at the feed-hole in the quilt. The queen cells need not be protected as the bees having younger queen cells of their own will readily accept them. Take care that they are not exposed to cold or crushed. Whilst being carried to the other hives they may be kept warm in a cardboard box in which is placed a small bottle of warm water.

All the original stocks may be expected to have virgin queens three or four days later, and these may be fertile and laying in little more than a week after that. Sometimes, of course, fertilisation is considerably delayed. The stocks headed by young queens will be unlikely to swarm during the season if given reasonable room.

When the young queens are safely laying each nucleus is provided with a full set of combs by the addition of full sheets of foundation and it and its parent colony then exchange positions. In this way the strength of the colonies will be approximately equalised.

The advantages to be gained from this method are the provision for the selection of stock, and the reduction of the queenless period by nine days in all the stocks save one, an important consideration at the beginning of the honey season.

5. NUCLEUS SWARMING.

Early in the season a nucleus colony, without queen, consisting of combs taken from a strong stock, as described on page 82, is given a ripe queen cell. When the resulting queen has been laying for a few days the colony is placed in a prepared hive and additional frames of brood (if available), empty combs, or frames of foundation are added to fill the hive. This is then placed on the stand of another strong stock at a time when the bees are flying, the latter being moved to a new position. The nucleus thus receives the field bees and becomes a strong stock.

A simple proceeding and one which does not involve the loss of field bees belonging to the nucleus is to place the hive containing the nucleus on the stand of the strong stock, and the latter on the stand vacated by the nucleus.

6. DIVISION AFTER NATURAL SWARMING.

Six or seven days after a natural swarm has issued from a hive the parent stock may be divided into nuclei, each having a queen cell. One of these nuclei should be left in the hive of the parent colony itself in order that it may receive returning field bees. Those removed should be strengthened by additional bees before they are removed to new positions in order to compensate for the loss of field bees which will return to the old stand. If however the division be made immediately after the parent colony has been deprived of its field bees as described in Chapter I there will be no need to strengthen any of the nuclei to make up for the loss of field bees. They can then be placed in any position without loss. The nuclei should be fed as they are built up and after the honey season they may be strengthened if necessary by additions of brood from the hive containing the parent stock.

7. DIVISION AFTER APPLICATION OF METHODS I AND I(A) (CHAPTER I).

The bees in the top box A may be divided into three nuclei and placed in new positions or left *in situ* after the second withdrawal of field bees. This must be done on the twelfth or thirteenth day, just before the young queens "hatch". At this stage the nuclei will take up new positions without loss. If box A is to remain in position on the original hive close fitting partitions must be inserted and suitable entrances provided.

I arrange for this in the following way:—

The box is placed so that the frames run from back to front. Two thin sheets of galvanised iron or three-ply wood form the partitions and are let into saw cuts or grooves on the inside of the box.

Two $\frac{5}{8}''$ holes are cut in the hive side walls, from which the young queens of the side nuclei fly to mate. The middle nucleus uses the back entrance (No. 5) in the screen board.

As the nuclei have ultimately to be removed, if so much increase is desired, it is easier to place them on their new stands when making the divisions.

The nuclei formed in this way must be strengthened with a little brood from time to time, and fed with syrup if they are to make strong stocks before winter comes.

8. CONTINUED INCREASE.

The following interesting method of rapid increase has been successfully practised for some years by Mr. H. A. Noyes, of Washford. It is based on a practice described by Dr. Miller in his "Fifty years amongst the Bees" and later published in the *British Bee Journal*, Vol. 47, p. 45.

The object of the method is to make a large increase of colonies in one season, breeding all the necessary young queens from one selected stock. To make it effective a minimum of eight or ten stocks is necessary and, except in an unusually good season, provision must be made for liberal feeding.

The procedure is as follows:—

Select the two stocks in the apiary and call their stands A and B, it being understood that the stock on A contains the queen from which it is wished to breed the young queens.

On a fine day, about noon, when bees are flying well, remove from stock A one frame of brood and bees and also the queen, and place them in a prepared empty hive. Put this hive on stand A in the place of the original stock.

Remove the original stock (now queenless) and put it on the stand of B after first removing stock B to a new position in the apiary.

On stand A there will now be the breeding queen with one frame of brood and bees. This hive will receive the flying bees of the stock originally there.

On stand B there will be a queenless stock receiving the flying bees from the stock originally on the stand. This stock will raise queen cells.

Now proceed to the other stocks in the apiary and take from them what frames of brood they can spare, but no bees, and fill up with foundation. Put all this brood into the hive on stand A. This hive will soon contain a very strong stock.

On the ninth day afterwards divide the stock B into nuclei so that each has one or more queen cells, and place these nuclei in new positions in the apiary.

The hive at B will now be empty. Take it or another empty hive to A and put into it the queen and a frame of brood with bees taken from A, as before. Place it on the stand of A, and remove the main stock, containing many brood combs, from A to B, as before.

Now go to the other main stocks in the apiary and again take what brood they can spare. Add all this brood to the nucleus left on A, as before.

In nine days' time again divide the stock on B into nuclei and remove them to new positions.

For a limited period these operations may be repeated every nine days. Food must be provided as necessary, especially for the building up of the newly-made stocks. At no time must queen cells be put into or allowed in the stock at A.

Considerable judgment is needed in taking brood from the other main stocks. These must not be drastically robbed, and on each occasion care must be taken to leave in each hive sufficient brood of all ages to ensure the maintenance of the stock.

CHAPTER VI

UTILISATION OF NATURAL SWARMS

THE time-honoured custom of throwing a swarm into an empty hive, or at the best into one fitted with comb foundation, and so leaving it without further attention until supers are needed, is now observed only by the thoughtless beekeeper.

The enlightened beekeeper reinforces a swarm so as to minimise the gradual but serious decrease in numbers resulting from the deaths of old bees which are not succeeded by young ones until the first brood "hatches", and he does so in such a way as to prevent the issue of after-swarms from the parent stock. He arranges also that the swarm shall commence work in supers at once rather than wait until a large brood nest has been filled.

It is to be remembered that the bees of a swarm, if left to nature, at first build combs of worker cells. If the queen is prolific she will deposit eggs in these cells almost as fast as the bees construct them, until finally the whole brood chamber will be almost filled with worker combs. If however the queen is old or otherwise inferior she does not keep pace with the comb-builders and the latter then begin to construct drone cells.

The value of combs for brood purposes is directly proportional to the number of worker cells they contain, and generally worker combs are ensured by hiving swarms on sheets of worker sized foundation. This however involves certain disadvantages. A newly-hived swarm quickly draws out the foundation and begins to fill the cells with honey. In so doing they store the honey where it is not needed and restrict the space in which the queen should lay. For these reasons some beekeepers consider that the cost of the first half dozen sheets of foundation is wasted.

We may now consider how we may ensure the best results from a prime swarm.

Hiving on Starters.

The swarm having been captured during the day, kept in shade, and cooled by a little water sprinkled over its temporary hive to prevent it from absconding, is transferred in the evening to a prepared hive placed near the stand of the parent colony which is subsequently moved to a new position. This prepared hive is filled with not more than six frames containing comb guides or "starters", i.e. strips of foundation about $\frac{1}{2}''$ deep. The remainder of the brood chamber is filled temporarily with dummies, some on one side of the frames and some on the other. The frames should be wired in the usual way. A strong swarm may be expected to build six combs entirely of worker cells. If it were provided with ten starters however considerable portions of some of the resulting combs would probably consist of drone cells. The brood nest is completed a fortnight later by the removal of the dummies and the insertion of additional full sheets of foundation.

Above the six frames containing starters is placed an excluder, and above this the supers taken from the parent stock.

The queen will now deposit eggs in the cells as they are constructed, and honey which is brought in will be stored in the supers instead of in the brood chamber. Thus there will be a gain of two or three weeks in respect of storage in the supers as well as a saving in the cost of comb foundation.

Reinforcement by Heddon Method.

Whether a swarm be hived on starters, full sheets of foundation, or empty combs, it is desirable to reinforce it, as well as to prevent after-swarms from the parent stock.

The second swarm usually issues from the parent stock on the eighth day after the first swarm. If left undisturbed there may be further swarms during the ensuing week. If the parent stock be deprived of its field bees on the seventh day "after-swarming" is usually prevented. This is accomplished in the following way:—

Place the newly-hived swarm by the side of the parent stock with the entrances on the same level. On the seventh day remove the parent stock to a new position some yards distant. All its flying bees will join and strengthen the swarm and the parent stock will give up the idea of further swarming.

A BETTER METHOD OF REINFORCEMENT.

The swarm is hived on a new stand by the side of the parent hive. A day or two later shake all the bees from the combs of the parent stock and let them run in to join swarm. Break down all queen cells and place the entire brood nest over a weaker colony. Examine two days later to make sure no queen cell has been missed.

By this method the swarm is strengthened to the greatest possible extent and any weak colony which receives the brood quickly becomes strong and ready for the honey flow.

UNITING SWARM AND PARENT STOCK.

Place the parent stock and swarm side by side. Break down all cells in the parent stock on the first, fifth, and tenth days so that no young queen can be raised. On the tenth day unite the parent stock to the swarm by placing it over the excluder as a super. As the remaining brood "hatches" the combs will be rapidly stored with honey and a heavy yield may be expected.

If it is desired to provide for requeening the swarm later on a three frame nucleus with a queen cell may be taken from the parent stock before the queen cells are destroyed.

FURTHER OBSERVATIONS.

Bees can be united without special precautions when honey is coming in. It is then safe to place a stock on the stand of another to receive its flying bees. Similarly a brood box containing bees may be placed bodily upon another, whereas in times of scarcity it is necessary to separate them temporarily by means of a sheet of newspaper which will be gradually nibbled away.

If two swarms issue at the same time and unite it is better to hive them as one swarm and give them extra super room than to attempt to separate them.

In an apiary where young queens are to be mated it is important that one or more of the best stocks have an abundance of drones. This can be ensured by inserting in the brood nests of the selected stocks one or two frames fitted with starters. These frames will soon be completely filled with drone brood. The frames must be inserted at least five weeks before the drones are likely to be needed for mating.

Should a stock become queenless and develop fertile workers it is useless to attempt to re-queen it. It can be united to another stock, or to a nucleus with brood and a fertile queen.

When the requeening of a swarm becomes necessary after the honey season any one of numerous methods of introduction described in text-books may be used. If the new queen has been purchased she will arrive in a travelling and introducing cage, usually with directions for her introduction. As these directions are given in all text-books it is unnecessary to repeat them here.

If the new queen has been bred in the apiary she may be introduced by a cage method or by one of the following direct methods:—

1. SIMMINS' DIRECT METHOD.

Remove the old queen during the early afternoon. In the evening, after dark, place the new queen in a safety matchbox by herself and keep her without food in a warm place,—e.g. the waistcoat pocket,—for 30 minutes. Raise the corner of the quilt, drive the bees back with a small puff of smoke, and let the queen run down between the combs. Cover up without further disturbance or delay, and do not examine the hive for a week.

2. THE WATER METHOD.

This method, devised by me in 1911, is specially useful when a laying queen is substituted for another laying queen. It is not advisable to use it for queens that have ceased laying for some days,—e.g. those imported by post. It is quite reliable when the queen of a nucleus is substituted for that of a stock, both having brood.

Proceed as follows:—

Remove the old queen. Place the new queen in an empty matchbox for five minutes. Then pour tepid water into the box and shake it very gently for a few seconds. Push the box open with the finger and allow the queen to walk down amongst the combs. Cover the hive immediately and do not disturb it for a week.

The great advantage of this method is that it saves a second visit to the apiary.

It should be remembered that it is risky to introduce a queen by any method if robbers are visiting the hive, and that the second day (the day after dequeening) is the most unfavourable for the liberation of a new queen however introduced.

CHAPTER VII

THE CAUSES OF SWARMING

FEW beekeepers can take part in the interesting work of swarm control without speculating on the cause of swarming. Many theories have from time to time been propounded to explain it but up to the present there is no general agreement on the matter.

The most widely accepted theory is that of the German investigator Gerstung who considered that the desire to swarm is occasioned by the presence in the hive of an excess of nurse bees and a consequent superabundance of larval food. Such a condition is reached when the queen has passed her peak period of laying or when she is limited for space in which to lay, whilst at the same time young bees are reaching the nursing age in increasing numbers.

Much evidence can be adduced in support of this theory from the conditions and behaviour of bees about to swarm and this evidence has been ably reviewed by Mr. D. M. T. Morland (*Annals of Applied Biology*, February, 1930).

Miss A. D. Betts, (*Bee World*, December, 1919), suggests that the bees themselves partake of the superfluity of larval food produced by the nurses and that this causes the desire to swarm. She adds the rather attractive suggestion that the absorption of the larval food or "royal jelly" by the workers may stimulate in them the dormant instinct of the primitive female bee, inducing in the nurse bees an instinctive memory of the brood rearing habits of their ancestors, and impelling the older bees to attempt pseudo mating flights; that when the bees issue as a swarm they do so

under a mating-flight impulse and in a state of helio-
tropism after a feast on the larval food.

The Gerstung theory is supported by the fact that
the addition of unsealed brood to a colony, which
gives increased employment to the nurse bees, delays
or prevents swarming.

On the other hand G. S. Demuth, the Editor of
Gleanings, disputes the theory on the ground that he
has proved that the substitution of sealed for unsealed brood, which must greatly increase the excess
of nurse bees in relation to their duties, is also effective
in preventing swarming, even when queen cells have
been started (*American Bee Journal*, September 1931).
Mr. Demuth's views deserve the greatest respect, but
it may be that his experience is the "exception which
proves the rule". The substitution of sealed for unsealed brood quickly results in greatly increased laying
room for the queen, and if the bees have any instinctive
foresight, as we may reasonably assume they have,
they may realize, when new larvae begin to appear
four days later, and when the decision to raise queen
cells or not must be made, that greatly increased
demands for larval food will soon be made upon them,
and that the nurse bees will then become fully
employed (vide p. 103).

Mr. W. Hamilton (*Bee World*, January 1932), after
producing various arguments to confute the Gerstung
and other theories suggests that swarming is caused
by a diminution of the characteristic odour of the
queen consequent on her receiving less larval food from
the nurses as the available space for her egg-laying
becomes constricted. Against this theory however is
the fact that some strains of Dutch and other bees will
swarm however much room be made available for the
queen.

The Gerstung theory by itself does not appear
directly consistent with certain conditions which deter

bees from swarming,—e.g., the addition of supers of drawn combs, or the advent of a honey flow. Mr. A. E. Cole (*American Bee Journal*, May 1925), after emphatically denying that an excess of nurse bees brings on the swarming impulse, makes the interesting and important suggestion that swarming may be caused by the presence in the brood nest of an excessive number of young bees which have reached the wax-secreting, comb-building, and honey-ripening stages, and for which there is insufficient employment. He considers that swarming is delayed, and reduced to a minimum, if these bees are withdrawn from the brood nest by the use of large hives and the timely provision of drawn combs in the supers.

Demuth (*Gleanings*, June 1930) and others insist that swarming is caused solely by the congestion of the brood nest. Congestion however would seem to be the natural consequence of an excess of unemployed young bees. They would crowd the brood nest looking for work, and especially would they concentrate there during cool nights or when a temporary cessation of nectar-gathering still further restricted their employ-ment. Congestion of the brood nest should therefore be regarded as an accompanying circumstance of the cause of swarming rather than as the cause itself. When the congestion is relieved in order to prevent swarming it is done by giving the nurse bees more brood to feed, the wax-secreters more comb-building and cell-capping to do, and the honey ripeners more storage room.

Although the young bees are most active as nurses during the first fortnight of their lives, and as wax-secreters during their third week, there is no possibility of testing the effect of completely withdrawing the latter from a stock because of the over-lapping of the brood-food and wax-secreting functions in respect to the ages of the bees. The attraction of bees of wax-secreting age to newly-added combs in the supers may therefore to some extent reduce an excess of nurse bees in the brood nest, and the possibility of discouraging swarming by this means tends to confirm rather than to detract from the Gerstung theory.

Strong support for the Gerstung theory, if this be extended to apply to the wax-secreting as well as to the nurse bees, comes from observation of marked bees of different ages composing a swarm. Such observations have been made by Morland (Rothamsted), Haydak (U.S.A.), and Taranov and others (Russia). Generally their results agree and show that a very large proportion of the bees of a swarm are of ages four to twenty days—that is to say, young bees in which the glands secreting brood-food and those secreting wax are in a state of maximum activity. These bees also have some household duties including the receiving of nectar and its inversion into honey, for which purpose they secrete enzymes. These are the swarm bees, called by Haydak "the jobless bees", and are accounted for thus:—

At the approach of the honey season the queen, who will have been laying fast for some months, slows up in her laying either on account of age or for want of cells in which to deposit her eggs. Consequently she needs and receives less food, and the amount of young brood to be fed diminishes. At the same time the amount of emerging brood is increasing to a maximum and soon there is a surplus of bees of the nursing and wax-secreting ages whose functions are no longer needed. These bees congregate on the combs and in spaces near the brood-nest. In some of them the ovaries become enlarged, causing them to become potential laying workers. They construct incipient queen cells and harass and guide the queen until she lays in them. Otherwise they continue in enforced idleness until the first queen cells are sealed. Continuity of the stock being thus assured, they issue as a swarm, accompanied by a few younger and older bees, and compelling the queen to go with them.

From these considerations it is logical to conclude that if we can so manage our bees that the nurses, wax-secreters, and nectar inverters—one or all of these, since their functions overlap—are kept fully employed we shall have no swarm, and experience confirms this. Thus our swarm-control methods include such expedients as:—

(1) Increasing the capacity of the brood box,

(2) Substituting foundation for old combs,

(3) Removing brood to supers, as in Demaree methods,

(4) Increasing the capacity of supers,

(5) Adding brood, sealed or unsealed, to the stock,

(6) Abstracting a nucleus and filling up with foundation,

(7) Separating flying bees from those which have not yet flown,

(8) Substituting a young queen for an old one,

(9) Relying upon the incidence of a nectar flow.

APPENDIX

METHOD I (pp. 18-35)

This method was devised for the prevention of swarming. Its use for queen rearing is incidental and secondary. In cold weather with lack of income the bees in Box A may construct poor queen-cells or may refrain altogether from making them. Experience has shown that the best queens are raised when bees of all ages are present, and income, especially of water, is continuous.

Mr. J. F. Bramwell, of Exmouth, has given much attention to the quality of the queens raised. By modification of the procedure in the case of one stock he secures from it all the queen cells he needs and these are raised under ideal conditions. He proceeds as follows—

As far as possible equalise the stocks during May by helping the weaker ones with brood so that they may all be ready for Method 1 at the same time. On the date chosen for the application of Method 1 put a frame of unwired foundation into the centre of the brood nest of the best stock. Apply Method 1 to all the other stocks. Six days later select one of the treated stocks for queen rearing. Break down all its queen cells in Box A. Remove two or three combs and substitute for them—

(1) A division-board feeder filled with diluted honey or syrup:
(2) The frame of foundation, now drawn out and containing eggs, from the best stock. (This comb may be divided into two portions by cutting horizontally through it so as to leave scolloped edges. The lower part may be fixed into another frame, scolloped edges downwards. Easily removable queen cells will be built along these edges.)

Now temporarily remove Box B to another stand and put Box A in its place under the supers. The stock will become very strong and will raise queen cells. When these are sealed (6 or 7 days later) distribute them to nuclei—deprived of their queen cells—made from the several Boxes A, or to Boxes A themselves after first destroying their queen cells. For greater security each queen cell may be suspended in a spiral cell protector. Only one queen cell should be allowed in each nucleus and in Box A of the cell producing hive. If weather conditions are good Boxes A and B of this hive may resume their normal positions, i.e. Box A above the screen-board and Box B below the excluder (vide pp. 34–35).

Nuclei and stocks raising queen-cells should be fed continuously.

METHOD I (A) (p. 46)

SCREEN-BOARD FOR W.B.C. HIVES

This is generally similar to that illustrated on p. 28 but is modified as follows—

The external dimensions of the board are the same as those of the lift which will rest upon it. The upper and lower battens fixed to the edges of the board are made sufficiently wide to enclose between them a space 15 inches square. Any W.B.C. brood box will cover this space. The wedge openings cut in the battens extend inwards to the central space and are covered by small sheets of tin to form tunnels.

When in use the board rests on the super and an outer lift rests on the outer edges of the board.

The level of the board may not coincide with the juncture of two lifts. In this case the space below the board may be closed either by constructing an additional lift of suitable depth or by pinning some material, such as canvas or sacking, around it.

The lifts may be secured against wind by means of wooden buttons fixed to the edges of the board.

IMPROVED SCREEN-BOARDS (p. 25)

Many alternatives to the loose wedges have been suggested. Amongst them we may commend (1) that of Messrs. Robert Lee; of Uxbridge, who substitute exterior slides of non-corrosive metal, and (2) that of Mr. H. L. Burt, of Midford, Bath,—made by Messrs. Burtt & Son, Gloucester, and E. H. Taylor of Welwyn— which is illustrated in the self-explanatory scale drawing below. The pivoted and separated wedges provide for variable openings, a stronger board, and easy manipulation.

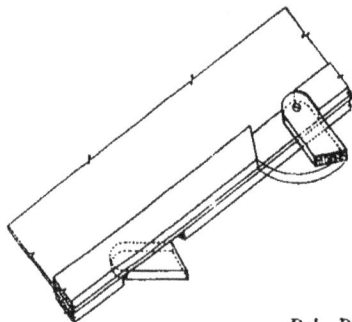

Del., R. M. Hopkins.

INDEX

Printed in Great Britain by Penwell Limited, Callington, Cornwall

www.ingramcontent.com/pod-product-compliance
Lightning Source LLC
Chambersburg PA
CBHW061928190326

41458CB00009B/2690